Global Knowledge Flows and Economic Development

OECD

ORGANISATION FOR ECONOMIC CO-OPERATION AND DEVELOPMENT

ORGANISATION FOR ECONOMIC CO-OPERATION AND DEVELOPMENT

Pursuant to Article 1 of the Convention signed in Paris on 14th December 1960, and which came into force on 30th September 1961, the Organisation for Economic Co-operation and Development (OECD) shall promote policies designed:

- to achieve the highest sustainable economic growth and employment and a rising standard of living in member countries, while maintaining financial stability, and thus to contribute to the development of the world economy;
- to contribute to sound economic expansion in member as well as non-member countries in the process of economic development; and
- to contribute to the expansion of world trade on a multilateral, non-discriminatory basis in accordance with international obligations.

The original member countries of the OECD are Austria, Belgium, Canada, Denmark, France, Germany, Greece, Iceland, Ireland, Italy, Luxembourg, the Netherlands, Norway, Portugal, Spain, Sweden, Switzerland, Turkey, the United Kingdom and the United States. The following countries became members subsequently through accession at the dates indicated hereafter: Japan (28th April 1964), Finland (28th January 1969), Australia (7th June 1971), New Zealand (29th May 1973), Mexico (18th May 1994), the Czech Republic (21st December 1995), Hungary (7th May 1996), Poland (22nd November 1996), Korea (12th December 1996) and the Slovak Republic (14th December 2000). The Commission of the European Communities takes part in the work of the OECD (Article 13 of the OECD Convention).

Publié en français sous le titre :

Les flux mondiaux de connaissances et le développement économique

Foreword

This book aims to provide recommendations to national and regional governments and development agencies on policies to stimulate the participation of firms and research organisations in global knowledge flows. It is based on a study undertaken by the Organisation for Economic Co-operation and Development (OECD) in collaboration with Scottish Enterprise, the principal economic development agency in Scotland, which made international comparisons of how governments and economic development agencies in various countries have been addressing this challenge. The study focused on policies and programmes to promote "knowledge-in" and "knowledge-out" through: i) promoting cross-border alliances, particularly by small and medium-sized enterprises, ii) stimulating knowledge transfer from Foreign Direct Investment ventures, and iii) attracting talented labour from overseas. Examination of these issues was placed within a broader conceptual framework that stressed the importance of building strong endogenous innovative capacity as a pre-requisite for successful participation in global knowledge flows.

Jonathan Potter of the OECD Local Economic and Employment Development (LEED) Programme designed and led the research project and edited this publication, working in collaboration with Martin Wight, Ross Brown and Michael Cannon from Scottish Enterprise. The members of the international review panel were: Professor Bjørn Asheim (University of Lund, Sweden), Mario Cervantes (OECD Directorate for Science Technology and Industry), Professor Philip Cooke (Cardiff University, United Kingdom), Professor John de la Mothe (University of Ottawa, Canada) and Paul Frater (Industry New Zealand, New Zealand).

This book is published under the responsibility of the Secretary-General of the OECD.

Table of contents

List of boxes

List of tables

List of figures

 GLOBAL KNOWLEDGE FLOWS AND ECONOMIC DEVELOPMENT – ISBN 92-64-10766-5 – © OECD 2004

ISBN 92-64-10766-5
Global Knowledge Flows and Economic Development

Chapter 1

Executive summary

by

Jonathan Potter (OECD LEED Programme)

This Executive Summary outlines the issues examined in the book, presents examples of learning model programmes and sets out the principal policy recommendations for areas seeking to participate in global knowledge flows.

Introduction

The key message of this book is that governments and economic development agencies should encourage domestic firms and research organisations to participate in global knowledge flows in order to underpin national and regional competitiveness.

The rationale stems from the growing importance of two key drivers of change in contemporary economic development. The first is the shift towards the "knowledge economy", in which economic development is strongly influenced by innovation performance, which depends increasingly on the generation, distribution and exploitation of knowledge. Networks for knowledge sharing are critical to innovation-led economic development and whilst many networks are local it is also important to exploit global connections. The second is globalisation, or the process of increasing interactions between countries through trade, investment, labour flows, strategic alliances and so on, facilitated by improvements in information and communication technologies (ICT) and political steps towards economic integration. Globalisation is creating new opportunities both for exploiting domestic knowledge overseas and for tapping into knowledge generated overseas. Thus globalisation and the shift to the knowledge economy both underpin innovation and competitiveness in modern economies.

Cross-border alliances among firms and research organisations, knowledge transfers from Foreign Direct Investment (FDI) and attraction of skilled labour and researchers can all contribute to the generation, transmission and exploitation of knowledge, thus increasing technological development in terms of potentially exploitable new products and services and new organisational methods.

Knowledge flows can occur in two directions: out of an area and into an area. Both can support domestic innovation and competitiveness. Knowledge flowing out of an area is often associated with an inflow of payments from licensing, fees and sales, and even where skilled people move overseas they often retain their local networks or return after a time with new knowledge and skills. Knowledge flowing into an area can be critical in enabling domestic firms and research organisations to access world class technologies, even though payments may need to be made overseas in order to secure knowledge inflows. Involvement in the processes of knowledge flows, both knowledge-in

 GLOBAL KNOWLEDGE FLOWS AND ECONOMIC DEVELOPMENT – ISBN 92-64-10766-5 – © OECD 2004

and knowledge-out, can accelerate innovation and economic development in countries and regions.

However, whilst there are benefits to be had from participation in global knowledge flows, there are also many barriers that public authorities can help address to ensure participation at an optimal level. These include lack of information about potential partners, language and cultural barriers, government regulations on labour mobility and inter-firm alliances, costs and difficulties of protecting intellectual property rights and difficulties obtaining specialist finance, as well as lack of capacity to adopt new technologies because of weaknesses in areas such as skilled labour, entrepreneurial vitality and networks.

Study background and issues

In 2001, Scotland introduced a new economic development strategy called "A Smart, Successful Scotland" committing the Scottish Executive (Scotland's government) and the Enterprise Networks (Scottish Enterprise and Highlands and Islands Enterprise) to deliver a "Global Connections Strategy" as part of a new overall economic development approach based on "growing businesses", "global connections" and "learning and skills". The spirit of the Global Connections Strategy is summarised in the following statement from the Scottish Minister for Enterprise and Lifelong Learning:

> "If Scotland is to thrive rather than simply survive in the rapidly evolving, knowledge driven, global economy we need to be more fully integrated within it: to be well connected physically, digitally and intellectually with the rest of the world." (From the Ministerial Foreword to "Scotland: A Global Connections Strategy".)

The Global Connections Strategy therefore intends to support the fullest possible Scottish participation in the global economy through: helping Scottish organisations increase their involvement in global markets, contributing to making Scotland a globally attractive location, attracting the most appropriate higher-value activities from overseas and encouraging more people to choose to visit, live and work in Scotland. These processes are critical to the capacity of any economy to generate high quality employment and entrepreneurship in a globalising knowledge economy.

This book provides international analysis and comparative examples to help clarify options for policy development in all countries and regions interested in the issue of stimulating cross-border knowledge flows. It focuses on three critical channels for global connections, namely *i)* cross-border alliances, *ii)* knowledge transfer from FDI, and *iii)* attraction of talented labour from overseas.

These issues are placed within a broader agenda that stresses the necessity of building strong endogenous innovative capacities before national or regional economies can create and sustain substantial knowledge-based global connections. The book therefore emphasises the importance of building innovative capacities in firms, people, research institutes and FDI affiliates and developing the necessary linkages between them to create vibrant national and regional innovation systems. By developing strong internal innovation capacities, economies will increase their attractiveness to foreign people and firms carrying knowledge as well as generate demand for global knowledge connections from domestic actors.

The study examined each of the three critical channels for global knowledge flows cited above. Key issues around each of them are briefly introduced below.

Cross-border alliances

Strategic alliances are increasingly important for producing the knowledge required to launch new products and services and new processes, particularly for small and medium-sized enterprises (SMEs) in knowledge-intensive sectors. These alliances may be with other SMEs, with large firms or with universities and research organisations. The importance of local and regional networking is already well recognised in work on national and regional innovation systems and clusters but what is less understood is that networks also need to be open to global sources of knowledge both to increase the scale of knowledge that potentially can be exploited and to counteract potential system "lock-in".

International networking by SMEs can be thought of as falling into two main types. Firstly, SMEs are often involved in horizontal co-operation with other SMEs or large firm competitors, for example for collaborative research, technology licensing or cross sales and marketing agreements. Secondly, they are often involved in vertical co-operation, where buyers and suppliers co-operate in product and process innovation or the outsourcing of specialised inputs. In both cases, joint ventures, mergers and acquisitions are often used as well as more informal agreements to gain access to relevant knowledge.

However, a number of barriers exist to cross-border SME networking. Firstly, there is a lack of information on overseas research sources and strategic alliance opportunities. Secondly, small firms often tend to have an inward-looking attitude to overseas linkages that prevent them from taking potentially profitable opportunities. Thirdly, SMEs may not be familiar with how they can use ICT to co-operate with firms overseas. Fourthly, internationalisation involves high costs and risks, especially for SMEs.

Various policy actions are therefore being undertaken in different countries to stimulate cross-border alliances including:

- Raising awareness of the importance of international networking and the potential business benefits.

 GLOBAL KNOWLEDGE FLOWS AND ECONOMIC DEVELOPMENT – ISBN 92-64-10766-5 – © OECD 2004

- Provision of information on potential overseas partners, for technology transfer and exporting.
- Facilitating contacts between local SMEs and potential overseas partners.
- Improving ICT links and encouraging their use by SMEs.
- Brokerage, trust-building and financial support for SME networking overseas.
- Organisation of joint overseas delegations, for example for firms in local enterprise clusters.
- Support to local firms to undertake the required research and development (R&D) and meet the required quality standards for partnership.
- Skills upgrading in the local labour force to facilitate the move into more knowledge-intensive networking.

This part of the study sought to examine a number of questions concerning the appropriate policy response, including:

- What sort of public programmes are operated by agencies in different countries to encourage international SME alliances? For example, local business development agencies commonly offer information and support to domestic firms seeking overseas alliances through one-stop shop systems, but what sort of more intensive programmes have been developed?
- How are these programmes being implemented? Does policy tend to focus on target groups of SMEs that are the most likely to internationalise? And what sort of agents deliver support?

Knowledge transfers from FDI

Theory and research show that FDI can have important wider dynamic impacts on the competitiveness of places by importing new knowledge and technologies into domestic firms and organisations. Foreign investors often have special firm-specific advantages, including efficient and leading operational practices and technologies, which can spin-off into host economies to drive productivity improvements and innovations in local firms and organisations.

A number of processes tend to be involved in inward knowledge transfers from FDI. Some of these processes involve active collaboration between foreign investors and local firms and institutions which might directly be encouraged by policy, for example:

- Networking through conferences, business forums, supplier groups, trade associations, etc.
- Visits and informal exchange of ideas with suppliers and customers on technical production issues, training, quality assurance, etc.

- Staff practices with suppliers and customers such as exchanges, secondments, mutually agreed recruitment and training.
- Joint projects with training providers and technology institutions.

Other processes involve more passive mechanisms such as the demonstration effect, the competitive spur, improved product inputs for customers and spin-out of trained staff and managers from foreign investors to local enterprises.

Governments and economic development agencies have developed a range of initiatives in order to support these processes, for example by encouraging foreign investors to increase local linkages and networking with local firms, to jointly develop new courses with training providers and to get involved in collaborative research with local institutions, often within the context of broader regional innovation strategies. Supplier development has been a particular focus of such initiatives.

The study therefore explored the following key questions for agencies seeking to maximise knowledge transfers from FDI:

- What sort of public programmes are operated by agencies in different countries to increase knowledge spillovers from FDI? What programmes exist to encourage foreign investors to increase the extent and quality of their local collaborations? What sorts of upgrading strategies have been developed for local firms to enable them to absorb potential knowledge spillovers?
- How are these programmes being implemented? For example, are knowledge transfer initiatives normally built into existing aftercare policies and at what scale should they take place: local, regional or national? Is it first necessary to overcome attitudinal and capacity problems in the domestic economy for these programmes to work?

Attracting talented labour from overseas

The cases of Silicon Valley and cities such as Toronto or Dublin demonstrate the important contribution that the attraction of talented overseas labour can make to the stimulation of entrepreneurship, often in knowledge-intensive sectors. Tolerance, social capital and a positive image appear to be key factors contributing to success in attracting skilled and entrepreneurial people. The quality of university systems is also clearly a critical asset for attracting overseas students and researchers. It must be recognised that many people move overseas temporarily as a career step. Their contribution to economic development can nevertheless be highly positive and, given suitable follow-on opportunities, people will often wish to stay. At the same time, it should be recognised that attraction programmes must fit with established immigration policies that seek to balance the interests of attracting skilled workers for economic development and protecting the jobs of domestic workers in occupations where there are labour surpluses. Indeed, the attraction of skilled and talented workers from overseas should go hand in hand with policies to upskill the domestic workforce.

Key questions addressed in this part of the study included:

- What sort of public programmes are operated by agencies in different countries for attracting talented people from overseas? For example, in certain countries and regions policy may have helped to build an attractive and credible image overseas that contributed to the attraction of skilled workers. In addition, policy may be able, directly or indirectly, to encourage domestic universities, colleges and enterprises to increase their intake of talented people from overseas. Programmes also exist in certain countries and regions (for example in parts of Ireland and Portugal) to encourage emigrants to return to set up businesses, to take skilled jobs or simply to invest in local businesses from their overseas residencies.
- How are these programmes being implemented? For example, are special measures needed to help integrate attracted skilled workers and entrepreneurs, for example in terms of recognition of qualifications, language course availability, visa regulations and so on?

Content of the book

Following this executive summary the five chapters of the review panel members are presented in turn, including reactions to the policies and programmes discussed in Scotland, comparisons between Scotland and other countries and recommendations for further policy development in Scotland and other countries. Each panel member tackles a different issue in their chapter. Bjørn Asheim focuses on SME innovation policy and regional innovation systems. Mario Cervantes examines the attraction, retention and mobilisation of highly-skilled labour. Philip Cooke looks at how to build knowledge-intensive sectors where a country or region has distinct and important assets and potential, taking the life sciences sector as a case study. John de la Mothe examines actions that can be taken at the city level and introduces the notion of constructed advantage. Paul Frater sets out the experience of the Pacific region in connecting to global knowledge flows. A concluding chapter pulls together the main points and draws out the overall policy recommendations and learning models.

Appendix 1 provides a description of the key programmes and issues surrounding the Scottish Enterprise Global Connections Strategy. Appendix 2 provides contact details for the expert panel and project leaders at OECD and Scottish Enterprise.

Learning models

The issue of global knowledge flows is an exciting one at this time as policy makers begin to realise their increasing importance to domestic competitiveness and to experiment with new initiatives to meet the challenges. There are many innovative programmes across the world that aim to promote connections to global knowledge flows and many models offering learning opportunities are described in this book. They include:

1. Supporting cross-border SME alliances:
   - The *TechAction Programme, Canada*. A community networking programme involving local firms and people in the cities of Ottawa and St Johns in preparing locally-designed strategies for entering the global knowledge economy. Such participative strategy building helps to tackle cultural issues around risk aversion and inward-looking mentality.
   - *New Zealand Diaspora Programmes*. "Kiwi Expatriates Abroad" and "World Class New Zealanders" provide networks enabling New Zealand businesses to make contacts with expatriates overseas in order to obtain a variety of help with their internationalisation efforts.
   - The *Scottish Global Companies Development Programme*. A programme providing internationalising SMEs with consultancy support for strategy development and action planning, as well as peer learning and specialist events.
2. Supporting knowledge transfer from FDI:
   - The *Scottish Alba Centre*. The Alba Centre provides a strong supporting environment bringing together growing domestic microelectronics design companies and foreign affiliates in the same sector. It includes a technology park campus, the Scottish Embedded Software Centre, the Virtual Component Exchange and the Institute for System Level Integration.
   - The *Singapore Semiconductor Industry*. Singapore has created a cluster of SME semiconductor service firms and domestic wafer fabricators on the back of foreign investment by stimulating close FDI-supplier linkages through the Local Industry Upgrading Programme and by liberalising immigration procedures for foreign engineers employed by foreign affiliates.
   - *ITRI, Taiwan*. Taiwan's Industrial Technology Research Institute (ITRI) has been very successful in forming research alliances with incoming foreign investors and commercialising outputs via domestic SMEs and firms on the Hsinchu technology park.
3. Attracting talented labour:
   - *Recruitment of Overseas Students in Canada*. Canada provides financial incentives for students registering on post-graduate programmes and is

GLOBAL KNOWLEDGE FLOWS AND ECONOMIC DEVELOPMENT – ISBN 92-64-10766-5 – © OECD 2004

doubling the number of awards of Masters and Doctoral Fellowships. Changes are also being made to immigration policies to facilitate the retention of international students.

- *Research Centre Funding in Ireland.* In Ireland two government-administered schemes will distribute about 1.2 billion euros to advanced research centres by 2007, through the Programme for Research in Third Level Institutions (PRTLI) and Science Foundation Ireland (SFI).
- *Talent Scotland.* An initiative to help Scottish firms hire senior electronic engineers quickly from overseas using a web tool providing information on job opportunities in Scotland and permitting engineers to post their CVs or directly apply for jobs.

4. Improving the capacity of endogenous innovation systems both to adopt knowledge from overseas and to generate knowledge that may be exploited overseas:

- *Finland's endogenous-based innovation systems.* Finland's innovation systems are built on key endogenous strengths in terms of good research and education, a competent workforce and good infrastructures, enabling Finland to attract and retain knowledge-based firms and researchers. This is supported by regional Centres of Expertise for research-industry networking and programmes to encourage university-SME links.
- *Baden Württemberg's Institutional Regional Innovation System, Germany.* Baden Württemberg's regional innovation system is underpinned by publicly-supported innovation centres that generate knowledge and encourage its exploitation, including 14 Fraunhofer Institutes and 300 Steinbeis Foundation transfer centres.
- The *Scottish Intermediate Technology Institutes (ITIs).* Three ITIs have been established in the fields of energy research, life sciences and communications technology in order to commission pre-competitive research that will lead to exploitable products. These institutes will help bridge the gap between basic university research and commercialisation of products.

Principal recommendations

The book also puts forward a number of recommendations for policy makers at national and regional levels. The following are among the major messages:

The stimulation of cross-border alliances by SMEs must work through two main approaches. Firstly, it is important to build a broader base of SMEs capable of generating and absorbing innovations. Effort is needed to encourage SMEs to "learn to innovate" and to upgrade national and regional innovation systems and the interactions within them. Secondly, it is important to assist that smaller subset of firms already at the stage of being ready to grow globally. SMEs are likely to face a range of barriers to global growth, including excessive time taken to reach a size and level of engagement where they have international recognition and credibility, excessive focus on incremental innovation rather than radical innovation, limited appreciation of international competitors and the nature of the international markets they are engaged in and inappropriate venture and development capital possibilities. Policy should assist companies with global growth potential to overcome these barriers and to successfully manage the major transitions they are likely to experience in their growth processes.

Foreign investors can also act as channels for importing overseas knowledge to host countries and regions. Explicit programmes should be built to improve networks and linkages between foreign investors and domestic firms and universities, particularly in key clusters. Foreign investors should be seen as among the key sources of knowledge within national and regional innovation systems and this potential should more effectively be tapped in terms of their contribution to upgrading domestic firms, participating in university research networks and exploiting commercialisation opportunities. More pro-active programmes to encourage the establishment of corporate spin-out firms should be considered.

To increase knowledge flows from the attraction of talented labour from overseas it is important to:

- Promote the development of well-paid job opportunities in key sectors/clusters through providing the right supporting environment for the growth of those clusters in terms of domestic skills, knowledge institutions, specialist finance, inter-firm networks and so on.
- Develop an attractive "people climate" by improving the perceived and actual quality of place.
- Use universities and research laboratories as levers to attract post-graduate students and researchers by ensuring they are attractive in terms of salary levels, quality of research infrastructure and quality of graduate education.
- Ease immigration rules and taxes to encourage the migration of highly-educated people working in knowledge-intensive industries.

There is also a wider need to develop strong domestic innovation systems as a pre-condition for connections to global knowledge flows. This requires plugging gaps in or enhancing the infrastructures, institutions and networks in the innovation system, recognising the interactions between initiatives. An important tool for this is the development of intermediary technology transfer agencies sitting between the universities and research laboratories undertaking basic research and firms commercialising the innovations. Such agencies help to improve the flow of knowledge within domestic innovation systems and to tap into knowledge from innovation systems overseas.

ISBN 92-64-10766-5
Global Knowledge Flows and Economic Development

Chapter 2

SME innovation policy and the formation of regional networked innovation systems

by

Bjørn T. Asheim (University of Lund, Sweden)

This chapter highlights the importance of building innovation systems to promote the involvement of SMEs in cross-border knowledge flows. Five main types of SME innovation policy tool are outlined: financial support for innovation projects and innovative new firms, technology centres, upgrading of regional innovation systems, innovation brokers and mobility schemes. Their relative importance in different contexts is discussed.

Introduction

Innovation has increasingly been recognised as the basis for promoting competitiveness by firms, regions and nations. This increased focus on innovation has placed innovation policy at the centre of policies for promoting regional and national economic development, and has put the question of how to best enhance innovation capabilities of firms, and in particular of SMEs, high on the research and policy agenda. Thus, this chapter will focus on innovation policies especially towards SMEs, as well as on strategies of building innovation systems. The arguments for this approach are to be found in the paper prepared by Scottish Enterprise included as Appendix 1 in this book. One of the key components that is viewed as contributing to closing Scotland's productivity gap and boosting the economic growth rate is:

- Increasing levels of innovation and R&D, particularly among indigenous business. This is among the highest priorities in transforming the Scottish economy and an area where the three channels for increasing international knowledge flows will play a crucial part. Scotland's total R&D expenditure as a proportion of GDP places it in the third quartile of OECD economies (compared to the UK's position in the 2nd quartile). Measures of research output from Scotland's universities (eg scientific research papers in leading journals per capita) suggests a high level of academic research activity, but this is not reflected in levels of R&D in the corporate sector.

This view of the importance of increasing levels of innovation and R&D to boost the economic growth rate is in accordance with the major shift in how theoreticians and policy makers look at the factors behind a country's long run economic performance, which has occurred during the last decades. The attention has increasingly come to be centred on the role of knowledge and innovation for firm, region and country performance. This has led to several insights, such as the systemic nature of innovation (Lundvall 1992, Edquist 1997), the role of the broader knowledge infrastructure for innovation processes in firms (Carlsson and Stankiewicz 1991), the potential of collaboration between industry, government and the higher education sector with a "triple helix" as the main model and metaphor for the relationship (Etzkowitz and Leydesdorff 2000), and the importance of institutional factors such as intellectual property rights, in fostering innovation and growth.

The systems approach to innovation is based on the interactive innovation model, which implies that innovation is viewed as a complex,

 GLOBAL KNOWLEDGE FLOWS AND ECONOMIC DEVELOPMENT – ISBN 92-64-10766-5 –

interactive, non-linear learning process inside firms and between firms and their environments. This view reflects a broad definition of innovation, to include both improvements in technology and better methods or ways of doing things. It can be manifested as new or changed products, services and production methods, new approaches to marketing, new forms of distribution and changes in management, work organisation and skills of the workforce (EC 1995). This view involves a critique of the linear, sequential model of innovation, which focuses exclusively on more radical, technological innovations. The interactivity of the innovation process refers to the internal collaboration between different departments of a company as well as to external co-operation with other firms (especially with customers and suppliers), knowledge providers, finance, training and public administration. Co-operation between local actors may in particular improve knowledge creation, informational exchange and learning which lead to innovation (Storper 1997). The broad understanding of innovation and the importance put on other types of knowledge than only the R&D-base imply an extension of the range of industries that can be viewed as innovative from typical high-tech industries, often located in central areas, to include also traditional, non-R&D-intensive industries often located in peripheral regions (Asheim and Isaksen 1997).

Although recognising that a strong competitive environment is the main stimulus for innovation in companies, the conceptualisation of innovation as interactive learning emphasises the importance of co-operation in innovation processes as well as a systemic view of innovation. The concept of an innovation *system* is, thus, based on the idea that the overall innovation performance of an economy depends, to a large extent, on how firms manage to utilise the experience and knowledge in other firms, R&D-institutes, the government sector etc. in innovation processes (Gregersen and Johnson 1997), and not just on the capability of every single firm (although competence and attitude by founders, leaders and workers are also of critical importance for their innovation capability). Firms combine internal and external resources and knowledge of many actors in building unique, firm-specific competencies that cannot rapidly be imitated by competitors (Maskell *et al.* 1998). Entrepreneurs and firms that fail to learn and change their products and ways of doing things will sooner or later exit from the market. A stronger emphasis on the system perspective seems quite relevant in the Scottish case, as Scottish Enterprise itself talks about the insufficiency of the traditional "hands-off" approach towards industrial policy in the UK as a whole, and the need for building a more innovative, ambitious and productive indigenous business base.

However, this recognition is not as explicit as it might be in the new policy initiatives expressed in the Scottish Enterprise "Global Connections Strategy", which appears to emphasise the role of non-indigenous actors to a greater extent than the endogenous based policy being implemented in most of the European countries in accordance with the view of innovation and innovation policies presented above. This is quite surprising as the most relevant comparative examples of successful development policy emphasise an endogenous approach to a much greater extent. For example, Ireland, which is one of the cases most often referred to when considering a successful innovation and economic growth policy, only took off on a sustainable growth path when the endogenous building up of human capital, especially in the area of ITC, became the corner stone of public policy in the software industry in the 1990s. Finland is one of the other success cases most often referred to and is perhaps the most significant example of an endogenous, top-down planned, systemic innovation policy. In order to emulate these widely acknowledged successes, Scottish Enterprise should emphasise new initiatives to ground their Global Connections Strategy in an endogenously focused, systemic, capacity-building approach.

Traditionally innovation policy in Finland has primarily been a top-down national policy with a very strong science and especially technology orientation. Typically, innovation policy in Finland is strongly embedded at the top governmental level through the Science and Technology Policy Council. This guarantees the legitimacy of the policy and underlines how important it is considered to be. It also ensures that the innovation policy initiatives are well coordinated and orchestrated between the various ministries within the government. Finnish policy makers see the industry-university relations as a crucial edge in global competition, and more innovative firms in Finland than in other European countries cooperate with universities. The Finnish cluster programme could be considered to be a new example of a public policy aiming at stimulating and supporting multilevel research cooperation.[1]

The Centres of Expertise programme represents the key policy initiative and the spearhead for promoting a more regional oriented development, and is based on knowledge and know-how of the highest level. This programme started in 1994, and the Ministry of the Interior is the main responsible authority. The overall objective of the Centres of Expertise is to identify regional strengths, and create economic growth by increasing the number of competitive products, services, enterprises and jobs based on the highest standard of knowledge and expertise. The Centres of Expertise programme is realised through co-operation between industry, local government, technology centres, universities, polytechnics, research institutes and other branches of public administration. A Centre of Expertise should aim at supporting specialisation and co-operation between regions, and increasing regional competitiveness. A main purpose of the Centres of Expertise programme is to bring leading experts in research, education and private enterprises in a

 GLOBAL KNOWLEDGE FLOWS AND ECONOMIC DEVELOPMENT – ISBN 92-64-10766-5 – © OECD 2004

region or network into close interaction. Seen in an international context the Centres of Expertise programme must be considered a very interesting organisational as well as institutional innovation aiming at supporting and stimulating regional development. It has also proved to be a very successful policy instrument.

Thus, a broadly defined innovation system based on good co-operation between the public and private sectors has long been a key concept in Finnish policy for knowledge-based development. Finland's present strengths are largely endogenous: the education and research systems, a competent workforce, good infrastructures – the basic factors for high-level knowledge and know-how. Finland's attractiveness in the global competition essentially rests on these factors.

Recent economic and societal development in Finland has essentially been based on the development of high technology, its effective utilisation and resulting increases in exports. This has led to a significant improvement of Finland's position in international competition. Thus, the most important lesson to be learned from recent developments is that success in innovation is a key to success overall. It makes for better business opportunities, the development of the public sector, enhanced productivity and higher employment levels. National innovation systems are at the heart of internationalisation as good collaboration between the public and private sectors is an intrinsic part of successful internationalisation strategies.

On a more general level, the internationalisation of the innovation system involves two challenges. On the one hand, the Finnish system must be able to compete for competent researchers and other research resources, projects and business research and development with other countries' systems and, on the other hand, Finnish players must be able to enter and make use of the opening markets. The main challenge for economic and societal development in conditions of growing global competition, is to be able to keep Finland sufficiently attractive to business and jobs and as a living environment in general.

Regional development is more and more clearly dependent on the way in which the region's own development factors evolve and how well they are made to interact both amongst themselves and with the national and international levels. Sustainable and balanced development rests on the enhancement of the region's own capacities and knowledge, such as basic information society skills. In this respect the situation differs from region to region. The smaller the region and the more specialised the knowledge, the more likely its need to find complementary knowledge, co-operation partners and larger user groups outside the region. New opportunities in the region are created by systemic collaboration between universities and polytechnics with a view to synergically complementary education provision. Especially for SMEs it is vital to have flexible access to the expertise of higher education institutions.

An interesting aspect of the role of universities in Finnish innovation policy is the fact that they have taken up the role of knowledge transfer organisations. Acknowledging that a major part of the knowledge needed in Finland is produced abroad, the Finnish government stresses the importance of international research co-operation, and the important role universities can play through co-operation with foreign universities and firms with regard to providing links between the regional and local economy and global networks. This could be an efficient strategy of promoting the internationalisation of local SMEs as well as providing access to global knowledge production and R&D for local SMEs.

The Finnish case demonstrates that in order to promote successful and productive connections between local and non-local factors the regional economy must have some "buzz", an endogenous milieu for knowledge based localised learning among locally embedded actors. This is necessary in order to be attractive enough for knowledge intensive FDI, as well as to be able to exploit the establishment of "pipelines" with external knowledge bases, which needs a sufficient local absorptive capacity, on the one hand, and internationalising local business on the other hand. Based on the Finnish case, the most efficient way to create such a local "buzz" is to establish an efficient innovation policy through the promotion of collaborative and systemic innovation support by building national and regional innovation systems. Thus, in the continuation of this chapter we will proceed by demonstrating how such co-operative and systemic innovation support can be brought about, with special reference to the needs of SMEs.

However, there are differences with regard to the specific type of market economy we find in Scotland compared to Finland and most other continental European countries, and this affects the discussion about the efficiency of various types of innovation policy concerning their ability and capacity to promote innovativeness and competitiveness. Soskice (1999) argues that different national institutional frameworks support different forms of economic activity. Co-ordinated market economies have their competitive advantage in diversified quality production, *i.e.* innovative activities based on interactive learning and knowledge accumulated collectively in the workforce, while liberal market economies are most competitive in industries characterised by radical innovative activities. Following Soskice, the Nordic and continental West-European economies can be referred to as co-ordinated market economies in contrast to what he calls uncoordinated or liberal market economies (*e.g.* the US and UK), where the main determinant is the degree of non-market co-ordination and co-operation that exists inside the business sphere and between private and public actors, as well as the degree to which labour remains "incorporated" in the labour market institutions and the financial system is able to supply long term finance (Soskice, 1999).

The role and characteristics of SMEs in the innovation process

SMEs have been increasingly recognised by policy makers as a target group for industrial and innovation policy following a re-emergence of the small firm sector in western industrialised countries since the mid 1970s. This is also the case in Scotland, where one of Scottish Enterprise's principal aims is to build "the capabilities and confidence of Scotland's SMEs to participate in overseas markets". This is considered to be "a crucial challenge for the future of the Scottish economy". However, to design innovation policy targeting SMEs demands insight into the role and distinctive characteristics of SMEs in wider production systems, as well as into important barriers SMEs face in enhancing their innovative potential. Thus, one of the key underlying aims of policy intervention is to help SMEs to overcome any size-related barriers that may limit their innovative capability (Smallbone *et al.* 2000). Therefore, in this part of the chapter, specific emphasis is put on what constitute good practice innovation policy principles, and how these can be adapted to different types of SMEs and territorial contexts. The chapter draws on comparative evaluations of almost 40 existing policy instruments for promoting the innovation capabilities of SMEs in eight European countries.[2]

Whilst there has been much debate about the relative contribution of firms of different sizes to innovation, a number of recent studies like the Community Innovation Survey reveal that small firms tend to make (technological) innovations less frequently than larger firms. However, it should be stressed that innovative activity, in the broad sense of innovation, is of potential importance to all firms, irrespective of size and sector, and that both large and small firms often play important, if different, roles in innovative activity. Small firms are often better at adapting, while big firms are better at producing major (radical) innovations. In addition dynamic complementarities often exist between large and small enterprises, particularly in the role of SMEs as suppliers and subcontractors to large firms in integrated production systems. It should also be stressed that small firms often make important contributions to the renewal of industries. Thus, new technology based small firms can have an important role in developing radically new innovation, while large firms typically innovate incrementally within existing technological trajectories. In this respect SMEs in particular can be important to high tech industries, as what ultimately become big innovations often start in small firms and end up being commercialised by big ones. According to Scottish Enterprise, the problem for Scotland in this context is that "no programme has explicitly been developed to improve the technological capabilities of local suppliers in Scotland".

Thus, innovation policies can also be based on the view that innovation is not necessarily – and not for the majority of SMEs – a product of investment in the knowledge producing sectors only, as seen in the linear innovation model. A major objective is to foster and speed up learning and innovation processes within firms, and between firms and their environment, where technology transfer may be one of the means (Nauwelaers *et al.* 1999). Such technology transfer could also take place between local firms and multinational corporations (MNCs) either abroad or locally as a result of inward investments. Until now "most foreign-owned subsidiaries have shown weak innovation tendencies, consequent of limited local R&D, and very restricted linkages into the rest of the economy" according to Scottish Enterprise. This has resulted in important technological inter-firm relationships being overlooked, which has reduced the scope for technology transfer to Scottish firms.

Regional innovation policy tools for SMEs

The general policy recommendations in these cases emphasise moves towards the support of networks and clusters of firms (that may have a regional, national or even a larger geographical extension), and the stimulation of interactive learning among firms and with knowledge organisations. The shift towards the interactive innovation model has, accordingly, increased the importance of the concepts of national and regional innovation systems in policy design. This perspective is also underlined in the OECD LEED methodology document for this study, which argues that "increased co-operation and networking between firms and organisations will be one of the keys to competitiveness within a globalising economy based on increasing international integration and flexible production. The importance of networking is already well recognised within national and regional innovation systems and clusters but networks also need to be open to global sources of knowledge and markets. Particularly for SMEs in knowledge-intensive sectors, strategic alliances are increasingly important for producing the knowledge required to launch new products and services or new processes. These alliances may be with other SMEs, with large firms or with universities and technological institutes."

Distinctive features of SMEs at the micro-level can have implications for their support needs. Thus, support directed at SMEs, including innovation support, needs to be based on an appreciation of three key size-related characteristics: *i)* their limited resource base, particularly with respect to finance and management resources, compared to larger firms; *ii)* the distinctive organisational culture, stemming from the combination of ownership and management that typifies the majority of SMEs; and *iii)* their lesser ability to shape and influence the external environment, meaning that smaller firms are typically faced with a more uncertain external environment than larger companies.

The heterogeneity that exists within the SME sector means that there is no single model or set of factors that adequately explains how and why innovation takes place. Nevertheless, the key role of internal factors on the nature and extent of innovative activity in SMEs has been emphasised by many researchers. These include both the personal characteristics of SME owners and managers, such as their background in terms of education and previous experience, and firm characteristics that include both resource and organisational issues, together with the interaction between the two. Although the key role of external orientation and networking has been heavily emphasised by many recent commentators, it is important to recognise that there can be disadvantages as well as advantages to such linkages and relationships in practice.

Barriers to innovation in SMEs

Problems constraining or preventing innovation are very diverse depending on type of SME and regional circumstances. Nevertheless, based on empirical investigations, six main innovation barriers related to the majority of SMEs have been identified (Kaufmann and Tödtling 2000). A first problem relates to *lack of funding*, *i.e.* problems in finding risk capital combined with a high risk of innovating in SMEs. Big firms have a broader set of products to rely on if *one* innovation project fails compared to small firms. Secondly, SMEs often have *limited management resources* (human capital), *i.e.* lack of personnel with higher education, and may, in addition, have problems in recruiting such personnel. A third barrier includes *limited technological competence* beyond experience based, tacit knowledge, at least if we exclude small technology based firms that often spin off from universities and research institutes.

It is important not to neglect types of problems other than financial and technical, which are constraining innovation although firms often do not sufficiently recognise them. Of central importance are strategic deficits and organisational weaknesses of SMEs. A frequent strategy deficit in the case of SMEs is a *narrow customer focus* making their innovation process dependent on their clients, whilst often SMEs also lack feedback from their clients. This may hamper market based product development, which may be important as SMEs have less power than big firms to influence the market and, thus, have to rely more on their flexibility. Another barrier relates to *lack of time* of managers. Daily work-overload of very few persons or even a single person in SMEs impedes or delays innovation projects. This may reflect barrier number two of problems in recruiting competent persons to perform innovation. However, it may also reflect the organisation of the work when the entrepreneur typically wants to overview every decision leaving little time for long-term innovation projects. A last barrier is *low strategic capability* in numbers of SMEs, *i.e.* not focusing on market research and innovation as an important part of the strategy of firms.

Main types of innovation policy tools for SMEs

The policy tools for promoting innovation activity in SMEs may be categorised into five main types according to the nature of the support (Table 2.1). To some extent the five main policy tools are directed towards lowering some of the above mentioned specific innovation barriers in SMEs. The five main types do not cover all possible innovation policy tools; nevertheless they reveal a portfolio of relevant tools.

Table 2.1. **Five main types of innovation policy tools**

Policy tool	SME innovation barriers targeted
1. Financial support for innovation projects and innovative new firms	Lack of risk capital
2. Technology centres providing services to regional firms	Lack of (codified) technological competence
3. Upgrading of regional innovation systems	Specific bottlenecks in the regional "system"
4. Innovation brokers for firms and knowledge organisations	Lack of technological competence, strategic planning, etc.
5. Mobility schemes to recruit specific kinds of personnel to SMEs	Difficulties in recruiting higher educated personnel

Source: Asheim and Isaksen, 2001.

The first and most frequently used tool gives financial support for innovation and R&D projects in new and existing firms. This kind of instrument in particular targets lack of risk capital for innovation activity in SMEs.[3] The second main type of instrument includes technology and knowledge centres that seem to be the most widespread tool for indirect support of technology diffusion and innovation in SMEs. The aim of the centres is to contribute to transferring technology to regional firms as well as to building networks between firms. The instrument is directed towards the problem of a limited resource base in many traditional SMEs, and technology centres constitute an external support infrastructure that SMEs can tap into.

A third main type of tool comprises the upgrading of regional innovation systems. This kind of tool looks beyond the mere innovation barriers in individual firms and is directed towards removing bottlenecks in, or strengthening, the total regional production and innovation system. Thus, this tool may be an important supplement to the other four main policy tools focusing more on individual firms. This is a particularly valid point for Scotland, as there seems to be a lack of system thinking in Scotland's industrial and innovation policy. A fourth main type is proactive innovation brokers. The brokers assist firms in identifying their innovation needs and in making contact with relevant knowledge organisations. This instrument may be directed towards several possible innovation barriers such as low technological competence, lack of market research and narrow strategic vision. Generally, SMEs often find it difficult to identify and articulate their own support needs, which highlights the potentially valuable role of brokers in offering diagnostic and evaluation support to SMEs. This instrument may be particularly relevant in stimulating firms with low innovative capacity to start a process of becoming more innovative.

The last main innovation policy tool is mobility schemes. These schemes aim at recruiting experts, *e.g.* a university candidate, to SMEs for a specific time period. The candidate works with a specific innovation project in the enterprise or else contributes to technology diffusion and strengthening the contact and co-operation between the firm and R&D institutes and higher education institutions. Some of the most effective means of promoting a demand for knowledge, and thus technology transfer, in SMEs involve strengthening the human resource base of the firm, *e.g.* by stimulating the employment of graduates in SMEs. Thus, mobility schemes are in particular directed towards the problem of recruiting higher educated persons to SMEs lacking technological competencies.

Classification of SME innovation policy tools

The five main innovation policy tools are different in several ways. In order to characterise and analyse the policy tools they are classified in two dimensions, resulting in a four quadrants table (Figure 2.1). The table distinguishes between two main aims of the support tools. Some tools aim at giving firms access to resources that they lack to carry out innovation projects, *i.e.* to increase the innovation capacity of firms by making the necessary resource inputs available, such as financial support for product development, help to contact relevant knowledge organisations or assistance in solving specific technological problems. The other type of instruments have a larger focus on learning, trying to change behaviour, such as the innovation strategy, management, mentality or the level of awareness in firms. Sharp divides do not exist between instruments on this dimension. To some extent all instruments may result in behavioural change in firms, however, some have as a more explicit aim to promote a more "innovative future" for firms.

Figure 2.1. **Two-dimensional classification of main innovation policy instruments**

	Aim of innovation support	
Target level of support	*Assign lacking resources to firms: Support the accomplishment of innovation projects*	*Learning to innovate: Change firms behavior*
Firm oriented	1. Financial support 4. Brokers **A**	5. Mobility schemes **B**
(Regional) system oriented	**C** 2. Technology centres	**D** 3. Upgrading of regional innovation systems

Source: Based on Nauwelaers and Wintjes, 2000, from Asheim and Isaksen, 2001.

Nevertheless, an appropriate way to design and implement an instrument aimed at assigning lacking resources to firms (following an evolutionary approach to policy) is to do it according to a learning-to-innovate framework. In line with this perspective the objective of policy instruments is not solely to provide scarce resources (such as financial assistance) to innovating firms *per se* but also to promote learning about R&D and innovation and the acquisition of new routines within firms. Lack of demand is often a bottleneck for financial incentives to innovation activity, *i.e.* that firms initially do not see the need to innovate, or alternatively, that firms do not have the capability to articulate their need for innovation. Thus, some policy instruments should also attempt to enhance demand for initial innovation activity of firms. This is clearly one of the ambitions of the newly proposed "Intermediate Technology Institutes" by Scottish Enterprise (see Appendix 1).

Providing incentives will usually include some learning by firms. However, incentives may not be sufficient since an effective diffusion of knowledge and routines about how to perform R&D and innovation is linked to cumulative and collective learning (learning by doing and by the doing of others). This requires, among other things, several innovation projects within firms to attain a critical mass, as well as an explicit codification and diffusion of relevant knowledge. Thus, in a learning perspective, financial incentives, for example, must include an explicit behavioural aspect. This approach is seen to best satisfy the needs of SMEs that have not seriously adopted innovation routines, and is looked upon as the right approach to introducing and diffusing R&D and innovation activity within firms. Thus, the ultimate target of the policy is to promote the endogenisation of innovation activity of enterprises, "*i.e.* to get enterprises, after a period of intense collective learning (and associated institutional change), to undertake an important share of these activities without or with very little government support" (Teubal 1997, 1166).

The other dimension includes the target group of instruments. Some tools focus on innovation and learning within firms, to lower the innovation barriers of firms, such as lack of capital or technological competence. Other instruments to a larger extent have regional production and innovation systems as their target group, aiming at achieving externalities or synergies from complementarities within the regions. The barriers may for example be lack of user-producer interaction or lack of relevant competence in the regional knowledge organisations to support innovation projects. In this context, Scottish Enterprise argues that there have been few attempts to develop links between inward investors and the Higher Education sector.

It is not straightforward to place the five main policy tools in one of the quadrants in Table 2.1, reflecting the fact that the policy types are crude classifications covering diverse individual instruments. Besides, every main policy type may include several different dimensions. This is in particular the case for innovation brokers. The brokers' task is to mediate contact between firms and relevant knowledge organisations that are able to support innovation projects. In this sense the instrument is of the A type in allocating inputs for innovation. The brokers may also stimulate long term contact between firms and knowledge milieus, *i.e.* strengthening the working of innovation systems (category C). The activity of brokers may also contribute to changing the innovative behaviour of firms in such a way that firms increase their external relations (category B and D).

The SMEPOL project found that policy tools were concentrated in category A in Figure 2.1, while there were few instruments in category D. Thus, this does not make Scotland that different from other comparable countries in Europe after all. However, no category of tools, be it B or D, is intrinsically better or more correct than another. The different types of instruments may also support each other. Successful instruments in category B mean, for example, that more SMEs become aware of and learn to innovate, which may increase the need for resources to support instruments in category A, such as more risk capital for firms that start new innovation projects when they have moved further on their learning curve. All types of instrument are relevant to different types of firm and different types of environment; however, the combination and importance of instruments may vary between different kinds of region.

To further analyse the relevant mix of innovation instruments in different kinds of regions, we need a more detailed understanding of each main type of policy tool. Thus, one or two instruments will be selected to illustrate important lessons and challenges for each of the main policy types followed by analysis of how to adapt instruments and support systems to a variety of contexts.

Examples of the use of the main SME innovation policy tools

Innovation Policy Tool 4) The TEFT technology attachés as brokers

This main type of policy tool differs from the other tools because of the active role of brokers in analysing firms and identifying innovation barriers in firms, which may or may not be recognised by the firms themselves. Innovation brokers may, above all, be relevant in stimulating less innovative firms to start a learning process to become more innovative, as well as promoting the formation of networks and interactive learning among firms and knowledge organisations. The newly proposed "Intermediate Technology Institutes" in Scotland could, at least in part, be looked upon as an example of a broker institution, even if it has a broader scope of commitments compared with the Norwegian TEFT programme.

The Norwegian TEFT program (Technology diffusion from research institutes to SMEs) may illustrate several strong elements of this policy type as regards raising SMEs' innovation capability.[4] TEFT started in 1994 and has a twofold goal structure. Non-R&D intensive SMEs with weak internal resources comprise the industrial target group of TEFT as this kind of firm is seen to be most in need of external R&D-competence in order to increase innovativeness. However, the target group also includes the five largest polytechnic R&D institutes in Norway, with the aim of changing their attitudes towards SMEs and strengthening their knowledge about SMEs' innovation needs.

Thus, TEFT's main aim is to increase nation-wide contacts and collaboration between SMEs and the five largest national research institutes in Norway, *i.e.* to draw non R&D-intensive SMEs into the national innovation system. Achieving this aim is the task of ten county-based *technology attachés* who function as brokers, or as organisers, animators or coaches in the innovation process of SMEs. Their task is to establish, through visits, the technological opportunities in SMEs that can best be met by the R&D activities of the five institutes within a specific technology project. Thus, TEFT is particularly concerned to lower barriers to co-operation between national R&D-institutes and SMEs, with the attachés acting as brokers, and in that way to encourage SMEs to use R&D institutes and to strengthen industry-science relationships.

TEFT has several strong components (Remøe 1999). TEFT's technology attachés can be seen as prototypes for a proactive working method in performing their roles as analysts of firms, brokers and mediators. The programme is able to mobilise less innovative SMEs to co-operate more with knowledge organisations and perform innovation projects. A potential weakness of the programme concerns whether TEFT's target group of firms always needs contact with the leading research milieus in Norway to increase their innovativeness. In SMEs with low R&D-levels, specialised consultants are

GLOBAL KNOWLEDGE FLOWS AND ECONOMIC DEVELOPMENT – ISBN 92-64-10766-5 – © OECD 2004

often able to do the job equally well as advanced research expertise. Co-operation with a R&D institution could, however, stimulate the appetite for more collaboration with R&D milieus – an important objective of the TEFT programme. According to the Finnish view on co-operation between SMEs and R&D institutions it is considered obvious that universities need to be more proactive also regarding knowledge users that are not involved in R&D activities through building up networks. However, the formation of such networks is a long-term process that requires active development of mutual interests and trust from both parties.

Experiences from TEFT point to some general challenges for innovation broker tools (Box 2.1). A first challenge is to ensure user-orientation, *i.e.* focusing on the needs of SMEs, when brokers have links to knowledge organisations that need project incomes. Another challenge is the strong dependence on the skills of the brokers, for example, in identifying needs and barriers that the firms themselves may not be aware of. Thus, the challenge is to professionalise what are often quite loosely defined jobs. One more challenge is to maintain a focus on innovation and avoid downgrading of the instrument to "basic" business development support. Innovation involves complex processes often simultaneously involving several activities in a firm, which make it difficult to focus on those activities most important to raise the innovation capability of SMEs. Finally, it may be a challenge to demonstrate the importance and impacts of "soft" tools such as the role of advisers and brokers, in connection with evaluations and in prioritising between different types of instrument.

Box 2.1. **General challenges for innovation brokers**

- Ensure user-orientation.
- Ensure strong broker skills.
- Focus on innovation needs and barriers.
- Analyse and demonstrate significance and impact.

Even if this Norwegian example is a domestic oriented policy initiative there should, in principle, be no problems in applying such a scheme in the international area. An illustration of this could be the way the Finnish government uses universities as international knowledge transfer organisations, *i.e.* that the universities as part of their "third task" could take on a broker role aiming at mediating knowledge from abroad (knowledge-in).

Innovation Policy Tool 5) Attracting Talented Labour to SMEs: Mobility schemes in Wallonia and Limburg

Mobility schemes contribute to recruiting educated and competent personnel to SMEs, lowering the barriers to SMEs employing people with a university degree and increasing contacts between firms and knowledge organisations. First-Enterprise ("Formation et Impulsion à la recherche Scientifique") in Wallonia, Belgium, illustrates challenges and lessons from mobility schemes.[5] The scheme, introduced in 1993, covers up to 80% of the salary for two years of a young researcher working part-time in an SME and part-time in a research laboratory. The researcher engages in a specific research project in the firm. Other eligibility criteria for access to a First project is that the researcher should spend enough time in the hosting research institute to allow a real knowledge transfer to the firms, and that the research team is competent in the relevant research field and motivated to engage in innovation projects in the firm.

A strong feature of First-Enterprise is the contact between firms and research laboratories through shared hiring of the young researchers. It may lead to more long-term links and co-operation with a research laboratory, generally increased use of knowledge organisations in innovation projects by the firms, and more long-term research based innovation in firms. A weakness on the supply side is the tendency by the research institute to see the instrument as a means of subsidising the salary of the researchers, while the firms' need for innovation support may recede into the background. Additionally, the instrument concentrates very much on the technological side of the innovation process, and does not, for example, deal with the commercialisation of innovations.

First-Enterprise points to several general challenges for mobility schemes (Box 2.2). The task of these instruments is to focus on the needs of the firms at the same time as R&D institutes are looking for opportunities to finance research activity. Furthermore, an important challenge is to achieve behavioural impacts within firms, *i.e.* to contribute to strengthening the innovation activity and changing the long-term innovative behaviour of firms. It is also a challenge to ensure a sufficient take-up of the schemes among several types of firms. In part this is also the ambition of the Scottish

Box 2.2. **General challenges for mobility schemes**

- Focus schemes on firm needs to strengthen innovation capability.
- Ensure an impact from the scheme on longer term behaviour.
- Ensure a sufficient take up of the scheme.

Intermediate Technology Institutes, which want to secure greater interchange between academia and business and industry. However, overall, the participation of SMEs in these schemes should result in their innovation activity and capability being strengthened.

In principle such schemes could also be used on an international level. A way of establishing links with the non-local, national and international competitive environment is to attract highly qualified people into the target country or region. When asking, as Scottish Enterprise does, if it is possible to change the perception or image of Scotland in advance of actual change, it seems that Scottish Enterprise has understood that it is not enough to attract firms: the "right" people also need to be attracted, for which mobility schemes could be a useful policy tool. Richard Florida (2000, 2002a, 2002b) calls for complementing policies for attracting firms with policies for attracting people, which means addressing issues of "people's climate" as well as of "business climate". Indeed, the former is seen as basic to the latter, in that the presence of human capital and talent is essential for attracting and developing high-tech industries and consequently for the economic growth of cities and regions. This suggests that the attention of politicians and planners should be more directed towards people, and not only towards companies, *i.e.* that strategies of business attraction are supplemented with a stronger attention to policies for talent attraction and quality of place (Florida, 2002b).

Innovation Policy Tool 2) Technological Institutes in Valencia and Emilia-Romagna

Technology centres seem to be the most common instrument to indirectly support technology transfer to and innovation in SMEs, and can thus be an efficient policy instrument to create local "buzz". Four Technological Institutes in Valencia, Spain, will illustrate the tasks and challenges for this policy tool.[6] The four institutes are sectorially specialised in one manufacturing industry: production of footwear, ceramics, textiles and toys respectively. These branches consist of numerous SMEs often agglomerated in specific industrial districts in Valencia.

The Technological Institutes were established in the 1980s. The most important task of the Institutes has been to assist Valencian SMEs in technological upgrading. The assistance consists of five main kinds of services adapted to important innovation barriers in different manufacturing sectors. First of all, the Institutes offer *information and documentation* through publication and information events related to matters of broad importance for regional SMEs. The Institutes inform and advise on design, fashion and CAD-CAM services. In addition, the Institutes perform *technical studies* of raw materials, components, production processes, machinery and finished products. Related to these studies are laboratory analyses to *test and certify* production and products to improve the quality of SMEs' activity and comply with international standards. A fourth task is *consulting* on technology transfer. The Institutes transfer knowledge and technology to SMEs in collaborative projects. Finally, the Institutes perform human resource *training* through specialised courses for technical personnel.

A strong point of the Valencian Technological Institutes is their embeddedness in local agglomerations of SMEs. The Institutes co-operate closely with, and have built up detailed knowledge about local SMEs, while at the same time being well connected to similar centres internationally. This is especially the case for the ceramic sector, which together with a similar Italian industrial district in Emilia-Romagna, is the world leader. Thus, there is a lot of local "buzz" as a basis for establishing international connections. Weak points are the low level of R&D-activity at the Institutes, the small number of joint R&D projects with local firms and little contact with R&D institutes. The use of more R&D competence in innovation activity is seen as a next necessary step to raise the competitiveness of Valencian SMEs. To be able to continue as important actors in the regional support system, the Institutes have to broaden their services from information, standard testing and training to include support of research projects in firms and to mediate research competence from other knowledge providers. Even though the Institutes have increased their self-funding they are seen to have less current relevance for firms because of a lack of R&D competence in the Institutes (Vazquez Barquero *et al.*, 1999).

Experiences from the Technological Institutes in Valencia point to an important general challenge for technology centres. The challenge consists of developing centres based on a "consultant" mode, transferring existing knowledge from their shelves to the firm, towards a "process consultant" mode (Nauwelaers and Wintjes 2000). Then the centres could work together with firms in their innovation projects, adapting their services to varying needs, for example by bringing in and perhaps modifying relevant knowledge taken from elsewhere.

The Italian industrial districts exemplify such strong local support of technology and innovation and the technology institutes in Emilia-Romagna provide a particularly good example. Enterprise support of a more proactive and systemic kind has been introduced recently by agreement amongst small firms, the regional government and the intermediary agencies. The period of conscious intervention accelerated during the 1990s when the ERVET (Ente Regionale per la Valorizzazione Economia del Territorio – the regional development agency) system of enterprise support (*i.e.* centres of real services) was put into place to assist the networked firms to deal with competition based on advanced technologies, on the one hand, and cheap overseas labour,

 GLOBAL KNOWLEDGE FLOWS AND ECONOMIC DEVELOPMENT – ISBN 92-64-10766-5 – © OECD 2004

on the other. Amongst these new initiatives are the following which must be considered of importance:

- *Technology Centres*: where smaller firms may access technology services; R&D information; professional training; high-quality communication services; management advice; marketing support.
- *Advanced Telecommunications*: optical fibre networks; International Standard Digital Network (ISDN); broadband services; specialist business services and shared facilities/costs; and video conferencing.
- *Access to Technology Foresight and Monitoring*: through networked information and services provision, including gateway services to international markets.

These policy initiatives have clearly assisted the SMEs in the regional economy to retain their international competitiveness as well as to be attractive enough for receiving knowledge intensive FDI in the form of acquisition of innovative SMEs but where the acquired operations remain in the region in order to further develop the knowledge base.

Box 2.3. **General challenges for technology centres**

- Develop from suppliers of standard services to "process consultants".
- Retain a public service mission (such as awareness-raising in SMEs) when increased self-funding is demanded.

Innovation Policy Tool 3) Upgrading of regional innovation systems: Large firm – supplier co-operation in Limburg

Instruments focusing on upgrading the innovation capability in networks or systems of firms may supplement the other main policy types, which concentrate on individual firms and the link between firms, technology centres and R&D institutes. The scheme "Knowledge intensive Industrial Clustering" (KIC) in Limburg, the Netherlands, exemplifies the third main policy type.[7] This instrument came into being in 1994 in a joint private-public initiative between the multinational copy machine producer OcÉ and the regional development organisation Syntens. The initiative started a pilot project in 1994 in order to upgrade the knowledge base of regional SME suppliers. Over the years the project has represented an influential model to other cluster policy projects in the Netherlands and forms the basis for a Dutch national "cluster policy". At present the main task of Syntens is to raise awareness among regional firms of the significance of continual innovation and competence building, as well as to provide all-round support for innovation.

The original project idea in Syntens was to give regional SMEs an opportunity to demonstrate their competence in product development and to learn from other local firms. As a regional innovation agency, Syntens considered the lack of innovative medium sized firms as well as the lack of communication and co-operation between some large companies and local SMEs to be a bottleneck for further industrial development in Limburg. OcÉ's motivation to join the project was to upgrade their supplies and to find new and more knowledgeable suppliers to contribute to the company's product development.

The first project formed seven so-called micro-clusters including 4 – 5 SMEs that co-operated in an innovation project. The firms started with a drawing or a lab model of new parts or modules for OcÉ's new colour copier to be developed into modules or systems that could be produced at marketable prices. The project, and mainly OcÉ, subsidised development costs in the region's SMEs. OcÉ found the project very successful and took over the organisation of the project from Syntens in 1996.

KIC has several strengths (Nauwelaers *et al.* 1999). The instrument stimulates learning, improves the innovation capability of SMEs as suppliers and subcontractors and encourages innovation co-operation among firms. KIC may be used alongside instruments supporting innovation projects in individual firms. Weak points relate in particular to the organisation of the project, such as vague and shifting procedures, the fact that few SMEs actually have the opportunity to engage in the project and the fact that SMEs do not participate in developing and running KIC. The original aim of the project, which was to upgrade the innovation capability of SMEs in the region in general, was lost as efforts became focused only on upgrading the suppliers of OcÉ. Thus, a general challenge for this kind of instrument is to avoid "lock-in" to the technological trajectory of the lead company and strong ties only to a small core of regional firms with few openings for new firms with supplementary knowledge and vision to enter the network. In KIC Limburg there is a need for other lead firms to join the project as well as for a supervising role from an intermediate partner like Syntens to avoid a situation where the project only contains a core of client firms and their suppliers and subcontractors. In Scotland such a role could be taken by the new Intermediate Technology Institutes, which would like to establish links with inward investors (FDI) in order to enhance linkages between incoming firms and universities and local research bodies.

Summing up

It is clearly important to make innovative small-firm growth and development a systematic part of the soft infrastructure available in Technology Centres and Parks. Another very instructive international case which could be referred to is the example of Linköping University, Sweden, with its membership-based Foundation for Small Business Development (SMIL) and Centre for Innovation and Entrepreneurship to help stimulate business growth through business development training, problem-solving groups and club or networking opportunities. So are the triangular links between university, Nokia and start-ups at the Oulu technology park, Finland, the monitoring of SME needs by the Technology Centres at Graz in Austria, and the proximity of Saab and Ericsson to the firms on the technology park at Linköping. Each of these successful examples of apparently secure interaction for furthering innovation by linking supply and demand for know-how market opportunities shows the importance of the network approach to innovative business growth and development.

So, putting together the main lessons from the cases, the following list covers many of the most important ingredients:

- Partnership amongst large, private firms, government, universities, intermediary agencies, research institutes and small firms (Triple-Helix model).
- Clear and transparent management which is flexible and open, not bureaucratic and hierarchical.
- Soft infrastructure of enterprise support for business development and management training for technology growth and innovation.
- Polycentric linkage to other key nodes in the innovation system locally and globally.
- Industrial variety, but not comprehensiveness, to reduce the dangers of monocentric dependence.
- Intelligence functions aimed at anticipating future needs and opportunities through technology foresight.
- Advanced telecommunications infrastructure to maximise economies of time and minimise transaction costs.
- Technology Centres to supply expert services for technology transfer from knowledge-centres such as universities and research institutes to small and large business enterprises and public organisations.

The customising of policy portfolios to Scotland's specificities

The above analyses of the main types of innovation policy offer some general lessons regarding the design of innovation support systems. First, there is a need for instruments in category B (*e.g.* mobility schemes) of Figure 2.1 to supplement the more traditional direct support schemes. The B-type of instruments (or A-type of instruments [*e.g.* brokers] designed and implemented according to the learning-to-innovate perspective) may broaden and renew the permanent core of already innovative firms that constitute the client base for several policy instruments. Tools of type B recruit new firms to the support system by focusing on changing the behavior of firms, *i.e.* raising the awareness, attitude and knowledge of innovation activity of firms. Secondly, more of these types of tools imply a need for receiver-oriented and proactive working methods in (part of) the support system in order to understand and respond to innovation needs in the target group of firms. Thirdly, the analyses also reveal a need for instruments of category C (*e.g.* technology centres) and D (*e.g.* RIS), which focus on stimulating interactive learning between firms and in co-operation with different kinds of knowledge organisations, and generally aim at lowering bottlenecks in regional innovation systems. C- and D-types of policy tool are urgent when we see "the main role of innovation policy, which aims to increase the capacity of a region and the capabilities of its SMEs to innovate, is to foster interactive learning within the firms and within the region" (Nauwelaers and Wintjes 2000, 200). Thus, some tools may focus on barriers in the working of regional production and innovation systems and not only on innovation barriers in individual firms. Finally, there is a need for instruments (or groups of instruments) to be all-round, *i.e.* including more than simply financial and technological support, as the innovation processes of SMEs are complex and involve both technological and non-technological needs.

However, there is no single permanent "best practice" policy, or mix of instruments, available for each and every situation, as regions and nations are very different. Thus, instruments and policy systems have to be context sensitive in being adapted to the needs and bottlenecks in different types of SMEs and regional circumstances. A proper sequence and mix of instruments is more appropriate than a search for universally and permanently adequate tools (*op. cit.*)

More specifically it could be argued that in regions containing numerous high technology firms and considerable spin-offs from knowledge organisations – a situation characterising mainly some central regions, which the Glasgow – Edinburgh corridor would most closely resemble in Scotland – instruments in category A and C in Figure 2.1 may be highly relevant. There, the target group is firms typically employing a relatively large share of higher educated personnel with R&D competence. However, the firms may at times be short of risk capital and links to potential clients nationally and internationally. In areas containing a small core of innovative firms and a large share of non or low innovative firms category B instruments, as well as innovation brokers, seem most appropriate rather than to provide ever more resources to the same group of innovative companies. This situation often typifies more peripheral areas with "thin" industrial milieus and few universities, R&D-institutes or other organisations to stimulate the innovation

 GLOBAL KNOWLEDGE FLOWS AND ECONOMIC DEVELOPMENT – ISBN 92-64-10766-5 – © OECD 2004

activity of SMEs, which in the Scottish context typically would be the Highlands and Islands. The B-type of instruments may raise the awareness of firms about the importance of innovation, teach firms to innovate and assist firms in making contact with relevant knowledge organisations outside the region. Firms may then develop into more real clients of the traditional financial support instruments in category A that supports specific innovation projects and assumes that firms are able to formulate innovation projects and applications. This scenario might perhaps be especially relevant to Scotland.

In areas with potential or working industrial clusters and R&D organisations, but with less interaction among local actors, instruments in category D may be relevant. In areas with several innovative firms, however, leading to small local ripple effects, instruments in category C and D are most relevant. This situation could be considered quite normal in Scotland with its lack of tradition of a systemic approach to innovation policies.

It could be argued that most of the general conclusions would be relevant also in the Scottish case partly strengthening already existing policy measures and partly introducing new and complementary tools primarily of the systemic and proactive types. A basic challenge to Scotland successfully implementing its new Global Connections Strategy aimed at internationalising Scottish SMEs and encouraging cross-border knowledge flows, is to create enough local "buzz" to be able to compete in the globalising learning economy. Thus, before trying to implement the new strategy careful consideration is needed of whether the prerequisites for being successful with the Global Connection Strategy are in place in Scotland in general as well as for the majority of SMEs specifically.

Networked territorial innovation systems

The lack of strong systemic and proactive elements in Scottish industrial and innovation policies represents an argument for further elaborating on the potentials of policy tools in category D, namely for supporting regional innovation systems (Figure 2.1). Previous research on regional innovation systems (Isaksen, 1999) has demonstrated that the innovative activity of firms to a large degree is based on localised resources such as a specialised labour market and labour force, subcontractor and supplier systems, a unique combination of different types of knowledge, local learning processes and spillover effects, local traditions for co-operation and entrepreneurial attitude, supporting agencies and organisations and the presence of important customers and users. Thus, governments and agencies at all spatial levels have increasingly become involved in seeking to stimulate innovation. At the regional level regional innovation systems and learning regions have been looked upon as a policy framework or model for implementation of long-term, development strategies initiating learning-based processes of innovation, change and improvement (Cooke *et al.*, 2000, Asheim, 2001, Asheim and Isaksen, 2002).

The formation of regional innovation systems (RIS) must be understood in this context of creating a policy framework aimed at a systemic promotion of localised learning processes in order to secure the innovativeness and competitive advantage of regional economies (Freeman, 1995, Cooke *et al.*, 2000). A regional innovation system involves co-operation in innovation activity between firms and knowledge creating and diffusing organisations, such as universities, colleges, training organisations, R&D-institutes, technology transfer agencies, business associations, and finance institutions. These organisations hold important competencies, train labour, provide necessary finance etc. to support regional innovation.[8] The different combinations of these elements can constitute environments with strong as well as weak systemic innovation potential (Table 2.2).

Table 2.2. **Characteristics of regions with strong and weak systemic innovation potential**

Strong RIS potential	Weak RIS potential
Infrastructural level	
Autonomous taxing and spending	Decentralised spending or taxation
Regionalised private finance	National private finance
Strategic infrastructure competence	Few infrastructure competences
Embedded universities/R&D labs	Disembedded universities/R&D labs
Organisational level: firms	
Workplace cooperation	Workplace antagonism
Externalisation	Internalisation
Innovation	Adaptation
Organisational level: policy	
Inclusive	Exclusive
Monitoring	Reacting
Consultation	Authorisation
Institutional level	
Consensus	Dissensus
Associative	Individualistic
Learning disposition	Introspective

Source: Asheim and Cooke, 1999.

Different types of regional innovation systems

It is important, analytically as well as politically, to distinguish between different types of regional innovation system. Thus, Asheim and Isaksen

(2002) distinguish between three main groups of RIS in order to capture some conceptual variety and empirical richness in this phenomenon, which resemble the typology of Cooke (1998). The first type may be denoted as *territorially embedded regional innovation networks* (Table 2.3), where firms base their innovation activity mainly on localised learning processes stimulated by geographical, social and cultural proximity without much interactions with knowledge organisations. The paradigmatic example of this type of RIS is the original industrial districts (Mark I) in the Third Italy.[9] This type is quite similar to what Cooke (1998) calls "grassroots RIS".

Table 2.3. **Some characteristics of three main types of regional innovation systems**

Main type of RIS	Location of knowledge organisations	Knowledge flow	Important stimulus for co-operation	Examples
Territorially embedded regional innovation network	Locally, however few relevant knowledge organisations	Interactive	Geographical, social and cultural proximity	Industrial districts (Mark I)
Regional networked innovation systems	Locally, a strengthening of (the cooperation with) knowledge organisations	Interactive	Planned, systemic networking	Industrial districts (Mark II), Graz, Oulu, Linköping
Regionalised national innovation systems	Mainly outside the region	More linear	Individuals with the same education and common experiences	Technopoles, science parks

Source: Modified from Asheim and Isaksen, 2002.

The innovation networks may be further developed into *regional networked innovation systems*. The firms and organisations are still embedded in a specific region and characterised by localised, interactive learning. However, the systems have a more planned character through the strengthening of the regional, institutional infrastructure, *i.e.* more R&D-institutes, vocational training organisations and other local organisations are involved in firms' innovation processes. The networked system is more or less regarded as the ideal-typical RIS; a regional cluster of firms surrounded by a local "supporting" institutional infrastructure. Cooke also calls this type "network RIS".

The third main type of RIS, *regionalised national innovation system*, is different from the two preceding in several ways. Firstly, parts of industry and the institutional infrastructure are more functionally integrated in national or international innovation systems, *i.e.* innovation activity to a larger extent takes place in co-operation with actors outside the region. Thus, this represents more of an *exogenous* development model of which science parks and technopoles are most representative. In general, science parks tend to have weak local co-operative environments (Henry *et al.*, 1995), which result in a failure to develop inter-firm networking and interactive learning in the parks, while technopoles are characterised by a limited degree of innovative interaction between firms in the poles, and by vertical subcontracting relationships with external firms (Asheim and Cooke, 1998). Secondly, the collaboration is to a larger extent based on the linear model, as the co-operation mainly involves specific innovation projects to develop more radical innovations and with the use of scientific, formal knowledge. Cooke (1998) describes this type as "dirigiste RIS".

The networked innovation system represents an *endogenous* development model as an attempt to increase innovation capacity and collaboration through public policy instruments. For SMEs in particular, to carry out more radical innovations there is often a need to supplement the informal, tacit knowledge with R&D-competence and more systematically accomplished basic research and development. In the long run most firms cannot rely only on localised learning, but must also have access to more universal, codified knowledge of, for example, national innovation systems. The creation of regionalised networked innovation systems through increased cooperation with local R&D-institutes, or establishment of some technology transfer agencies, centres for real services etc., may precisely give firms access to information and competence which may supplement local competence and, thus, increase the collective innovative capacity and counteract "lock-in" of, in particular, regional clusters of SMEs.

The networked regional innovation system is different from the embedded innovation network due to the *systemic* dimension of the former, which requires that the relationships between the elements of the system must involve a degree of long-term, stable interdependence. However, it is still an example of a bottom-up, interactive innovation model, and thus, represents an alternative to regionalised national innovation systems. The systemic, networked approach to regional innovation systems brings together regional governance mechanisms, universities, research institutes, technology transfer and training agencies, consultants and other firms acting in concert on innovation matters as well as promoting a socially interactive culture of co-operation (Asheim and Cooke 1999).

The networked regional innovation systems represent a planned interactive enterprise-support approach to innovation policy relying on close university-industry cooperation. Large and smaller firms establish network relationships with other firms, universities, research institutes, and government agencies. Examples of such networked innovation systems can either be found in regions in Germany, Austria and the Nordic countries, where this model has been the more typical to implement (Asheim and

Cooke 1999), or in later stages in the evolution of industrial districts, which were previously characterised by territorially embedded, innovation networks (*e.g.* industrial districts Mark II in Emilia-Romagna).

The strength of the network approach is based on the interactive innovation model, which points at co-operation and localised learning as key factors in promoting innovativeness and competitiveness. The idea was that firms learn most efficiently from other firms, and, thus, improved inter-firm co-operation was considered to be the best way to improve the innovative capabilities of regional SMEs. Indeed, many case studies have demonstrated the advantages of territorial agglomerations in stimulating interactive learning between universities and private companies. And finally, the planned, systemic elements found in many of these cases emphasise the potential for a fully developed regional innovation system to support localised learning (Cooke *et al.* 2000).

Concluding remarks: RIS and the impact of FDI on local firms' innovativeness and competitiveness

The importance of the regional level is confirmed by results from the European comparative cluster survey (Isaksen, forthcoming), which shows that regional resources and collaboration are of major importance in stimulating economic activity in clusters due to the key role of place-specific local resources and regional innovation systems. That study found that in many clusters, firms increasingly find relevant research activities and other supporting services inside the cluster boundaries (Isaksen, forthcoming). Isaksen found that this was supported by formal organisations and social institutions, which helped to co-ordinate activities and manage transactions in the clusters. However, the survey found an increased presence of MNCs in many clusters, and also that firms in the clusters increasingly source major components and perform assembly manufacturing outside of the clusters (Isaksen, forthcoming). Also Tödtling *et al.* (forthcoming) found support for clustering, because of the importance of social interaction, trust and local institutions. Yet they also noted that both local and distant networks are often needed for successful co-operative projects, in particular for projects of innovation and product development where it is usually necessary to combine both local and non-local skills and competences in order to move beyond the limits of the region (Tödtling *et al.*, forthcoming). Interactions and connections with non-local innovation systems thus serve to prevent path-dependency of local industry and the local network, which could culminate in "lock-in" situations demonstrating the "weakness of strong ties" (Grabher 1993). This risk is built into an innovation system strategy, as the key rationale precisely is to support the positive effects of economic path-dependency and industrial specialisation.

This confirms results from previous research which revealed that the regional level is neither always nor even normally sufficient for firms to remain innovative and competitive, and which pointed to the additional importance of innovation systems at the national and international level for firms in regional clusters. The Finnish view on this is that fierce global competition forces companies to search for the most advanced university institutes worldwide. The most successful companies have at least to engage in global innovation networks and to make use of foreign basic research. Thus, firms are in need of innovative co-operation and interaction with world-class, national and international competence centres and innovation systems in order to stay competitive. In the Finnish case it is argued that national and international networking by the regions must be intensified through collaboration between public and private players with a view both to utilising knowledge available elsewhere and, especially, to improving innovation services needed by small and medium sized enterprises.

This represents an example of a multi-level approach to innovation systems and knowledge infrastructures, *i.e.* that different forms of knowledge must be accessed at different parts of the knowledge infrastructure (*i.e.* functional levels) and at different spatial scales (*i.e.* territorial levels) (Cooke *et al.*, 2000, Freeman, 2002). These tendencies will have consequences for the relevant types and scales of innovation systems in order to accommodate challenges from globalisation and adapt and modify the competitive behaviour of local firms. This will especially challenge the future role of regional innovation systems (Asheim and Isaksen, 2002). One approach to an alternative institutional framework for innovation support in a context "where global value chains are integrating with regional clusters" (Cooke, 2001b, p. 7) could be what Cooke calls an "Industrial Innovation and Learning System" building on "both the national and regional innovation systems, thus integrating the multi-level governance concept, and [... which...] integrates the important 'learning system' dimension" (Cooke, 2001b, p. 7).

One way to integrate or link up regional clusters with global value chains is through FDI. However, the impact of FDI should be analysed within a framework of varieties of capitalism and business systems (Soskice 1999, Whitley 1992). The impact and consequences of incoming FDI seem to a large extent to be determined by the behavioural characteristics of the business systems of which FDI serve as carriers, through a strong influence of home-based institutions on the structure and behaviour of MNC, and which reflect differences among national political economies (Hall and Soskice 2001). Thus, the analytical focus should turn towards studying FDIs as business system interfaces. The business system perspective should be used to enrich the study of how multinationals function as knowledge infrastructures constituting learning interfaces between potentially divergent knowledge architectures of different foreign and domestic companies.

Placed in the context of Soskice's theoretical framework the systemic view on innovation is typically accompanied by the institutional framework of a co-ordinated market economy. Should Scotland then neglect all the policy learning and recommendations in this report, which have primarily been taken from countries characterised by coordinated market economies, and just adopt the new economy innovation system, or should it choose a more heterodox mix of policies trying to get the best out of each system? Could anything be learned from the Finnish case, even if Finland too is a typical representative of a co-ordinated market economy? Finland considers that its contribution to the development of the European research and innovation area could well be to highlight a holistic approach to the innovation system in the European debate and the resulting enhancement of European co-operation in the development of different innovation sectors. Thus, even if the UK, in addition to the US, is the main representative of a liberal market economy, and, thus, leaves Scotland with apparently few alternative options, Scotland does not in general have all the same problems and challenges as England due to differences in size, economic structure, history, etc. The devolution policy in the UK might also give Scotland a larger relative autonomy in the future which could be used to obtain more control over vital domestic affairs such as the area of industrial and innovation policies. This would imply bigger windows of opportunity for choosing a more independent path of development with the ambition of arriving at a policy that deals with Scottish problems and challenges in the most adequate and efficient way.

Notes

1. The information on Finland is taken from a publication from Science and Technology Policy Council of Finland: "Knowledge, innovation and internationalisation". Helsinki 2003.
2. This section is based on results from the European research project "SME policy and the regional dimension of innovation" (SMEPOL) financed by the Targeted Socio-Economic Programme (TSER) Programme (EU, 4th Framework Programme, 1997-1999). SMEPOL was a collaborative project including seven research institutions in Europe: University of Economics and Business Administration, Austria; University of Southern Denmark, Denmark; Universitá degli Studi di Pavia, Italy; Maastricht Economic Research Institute on Innovation and Technology, The Netherlands; The STEP Group, Norway (project coordinator); Universidad Autonoma de Madrid, Spain; and Middlesex University, The United Kingdom (Asheim *et al.* 2003; Asheim and Isaksen 2001).
3. As this policy tool is the most well-known in a UK context, it will not be dealt with in detail in what follows.
4. The analyses of TEFT are based on Remøe (1999).

5. The analysis of the instrument is based on Nauwelaers *et al.* (1999).
6. The analyses of the Valencian technological Institutes are based upon Vázquez Barquero *et al.* (1999).
7. The analyses of KIC build on Nauwelars *et al.* (1999).
8. This conceptualisation of regional innovation systems corresponds with the one found in Cooke *et al.* (2000). In their words any functioning regional innovation system consists of two sub-systems: *i)* the knowledge application and exploitation sub-system, principally occupied by firms with vertical supply-chain networks; and *ii)* the knowledge generation and diffusion sub-system, consisting mainly of public organisations.
9. Industrial districts Mark I are districts without public intervention, which still is the normal case in Veneto, while industrial districts Mark II are characterised by public intervention often on a private public partnership basis which is the case in Emilia Romagna.

References

Asheim, B. T. (2000), Industrial districts: The contributions of Marshall and beyond, in Clark, G., M. Feldman and M. Gertler (eds), *The Oxford Handbook of Economic Geography*, Oxford University Press, Oxford, 413-431.

Asheim, B. T. (2001), Learning regions as development coalitions: Partnership as governance in European workfare states? Concepts and Transformation, *International Journal of Action Research and Organizational Renewal*, 6, 1, 73-101.

Asheim, B. T. and A. Isaksen (1997), Location, agglomeration and innovation: towards regional innovation systems in Norway? *European Planning Studies*, 5, 3, 299-330.

Asheim, B. T. and P. Cooke (1998), Localised Innovation Networks in a Global Economy: A Comparative Analysis of Endogenous and Exogenous Regional Development Approaches. *Comparative Social Research*, Vol. 17, JAI Press, Stamford, CT – London 1998, 199-240.

Asheim, B. T. and P. Cooke (1999), Local learning and interactive innovation networks in a global economy. In: Malecki, E. and P. Oinäs (eds), *Making Connections*. Ashgate, Aldershot 1999, 145-178.

Asheim, B. T. and A. Isaksen (2001), Regional innovation policy towards SMEs: Learning good practice from European instruments. In: *The Regional Level of Implementations of Innovation Policies*, Conference Proceedings from a workshop in Brussels November 2000. Community Research, European Commission, EUR 19846, Brussels 2001, 73-94.

Asheim, B. T. and A. Isaksen (2002), Regional innovation systems: The integration of local "sticky" and global "ubiquitous" knowledge, *Journal of Technology Transfer*, 27, 77-86.

Asheim, B. *et al.* (eds) (2003), *Regional Innovation Policy for Small-Medium Enterprises*. Edward Elgar, Cheltenham, Glos.

Carlsson, B, and R. Stankiewicz (1991), On the Nature, Function and Composition of Technological Systems, *Journal of Evolutionary Economics* 1, 93-118.

Cooke, P. (1992), Regional innovation systems: Competitive regulation in the New Europe, *Geoforum*, 23, 365-382.

Cooke, P. (1996), Networking for Competitive Advantage, National Economic and Social Council, Dublin.

Cooke, P. (1998), Introduction: Origins of the concept, in Braczyk, H., P. Cooke and M. Heidenreich (eds), *Regional Innovation Systems*, UCL Press, London, 2-25.

Cooke, P. (2001a), Regional innovation systems, clusters, and the knowledge economy, *Industrial and Corporate Change*, 10, 4, 945-974.

Cooke, P. (2001b), Industrial Innovation and Learning Systems: Sector Strategies for Value Chain Linkage. Chapter 6 in UNIDO World Industrial Development Report (WIDR), Vienna.

Cooke, P., P. Boekholt and F. Tödtling (2000), *The Governance of Innovation in Europe. Regional perspectives on global competitiveness*, Pinter, London.

Edquist, C. (ed.) (1997), *Systems of Innovation: Technologies, Institutions and Organisations*. Pinter, London.

Etzkowitz, H. and L. Leydesdorff (2000), The dynamics of innovation: from National Systems and "Mode 2" to a Triple Helix of university-industry-government relations, *Research Policy*, 29, 109-123.

Florida, R. (2000), Place and the New Economy. Transcript of "Champions of Sustainability Lecture". Broadcast on WQED August 27, 2000.

Florida, R. (2002a), Bohemia and economic geography. *Journal of Economic Geography*, 2, 55-71.

Florida, R. (2002b), The Economic Geography of Talent. *Annals of the Association of American Geographers*, 92, 4, 743-755.

Freeman, C. (1987), *Technology policy and economic performance: Lessons from Japan*, Pinter, London.

Freeman, C. (1995), The national system of innovation in historical perspective, *Cambridge Journal of Economics*, 19, 1, 5-24.

Freeman, C. (2002), Continental, national and sub-national innovation systems – complementarity and economic growth. *Research Policy*, 31, 191-211.

Grabher, G. (1993), The weakness of strong ties: the lock-in of regional development in the Ruhr area, in Grabher, G. (ed.), *The embedded firm. On the socioeconomics of industrial networks*. Routledge, London and New York, 255-277.

Gregersen, B. and B. Johnson (1997), Learning Economies, Innovation Systems and European Integration. *Regional Studies*, 31, 5, 479-490.

Hall, P. A. and D. Soskice (2001), *Varieties of Capitalism. The Institutional Foundations of Comparative Advantage*. Oxford University Press, Oxford.

Henry, N., Massey, D. and Wield, D. (1995), "Along the Road: R&D, Society and Space", *Research Policy*, 24, 707-726.

Isaksen, A. (ed.) (1999), Regionale innovasjonssystemer. Innovasjon og læring i 10 regionale næringsmiljøer. STEP-report R-02, The STEP-group, Oslo.

Isaksen, A. (forthcoming), Regional clusters between local and non-local relations: A comparative European study, in Lagendijk, A. and P. Oinas (eds), *Proximity, distance and diversity: Issues on economic interaction and local development*, Ashgate, Aldershot.

Jones-Evans, D. and Klofsten, M. (1997), "Universities and Local Economic Development: The Case of Linköping", *European Planning Studies*, 5, 77-93.

Jussila, H. and Segerståhl, B. (1997), "Technology Centres as Business Environments in Small Cities", *European Planning Studies*, 5, 371-384.

Kaufmann, A. and F. Tödtling (2000), Innovation patterns of SMEs. In Isaksen, A. *et al.* (eds), SME policy and the regional dimension of innovation. Report to the European Commission (TSER Programme). STEP-group, Oslo (118-145).

Lundvall, B.-Å. (1992), *National Innovation Systems: Towards a theory of innovation and interactive learning*. Pinter, London.

Lundvall, B-Å. and S. Borrás (1997), The globalising learning economy: Implications for innovation policy. Report from DG XII, Commission of the European Union.

Maskell, P. *et al.* (1998), *Competitiveness, Localised Learning and Regional Development. Specialisation and prosperity in small open economies*. Routledge, London and New York.

Nauwelaers, C. *et al.* (1999), SME Policy and the Regional Dimension of Innovation: The cases of Wallonia and Limburg. MERIT, Maastricht. SMEPOL report No. 4.

Nauwelaers, C. and R. Wintjes (2000), Towards a new paradigm for innovation policy. In Isaksen, A. *et al.* (eds), SME policy and the regional dimension of innovation. Report to the European Commission (TSER Programme). STEP-group, Oslo (194-222).

Nelson, R. (ed.) (1993), *National Innovation Systems: A comparative analysis*, Oxford University Press, Oxford.

Remøe, S. O. (1999), TEFT: Diffusing technology from research institutes to SMEs. STEP working papers A-03-1999. STEP Group, Oslo.

Smallbone, D. *et al.* (2000), The role and characteristics of SMEs. In Isaksen, A. *et al.* (eds), SME policy and the regional dimension of innovation. Report to the European Commission (TSER Programme). STEP-group, Oslo (28-43).

Soskice, David (1999), Divergent production regimes – uncoordinated and coordinated market economies in the 1990's. In Kitchelt *et al.*, *Continuity and change in contemporary capitalism*. Cambridge University Press, 101-134.

Smith, K. (1997), Economic infrastructures and innovation systems, in Edquist, C. (ed.), *Systems of Innovation: Technologies, Institutions and Organisations*. Pinter, London, 86-106.

Storper, M. (1997), *The regional world. Territorial development in a global economy*. The Guilford Press, New York and London.

Teubal, M. (1997), A catalytic and evolutionary approach to horizontal technology policies (HTPs). *Research Policy*.

Tödtling, F., M. Trippl and H. Brathl (forthcoming), Networking and project organisation in the automotive industry: the case of Styria, in Lagendijk, A. and P. Oinas (eds), *Proximity, distance and diversity: Issues on economic interaction and local development*, Ashgate, Aldershot.

Vázquez Barquero, A. *et al.* (1999), SME policy and the regional dimension of innovation: The Spanish report. GIDIT, Departamento de Estructura Económica y Economía del Desarolla. Facultad de Ciencias Económica y Empresarialas, Universidad Autónoma de Madrid. SMEPOL report No. 6.

Whitley, R. (1992), The comparative study of business systems in Europe: Issues and choices, in Whitley, R. (ed.), *European business systems: Firms and markets in their national contexts*. Sage Publications, London, 267-284.

ISBN 92-64-10766-5
Global Knowledge Flows and Economic Development

Chapter 3

Attracting, retaining and mobilising high skilled labour

by

Mario Cervantes
(Directorate for Science, Technology and Industry, OECD)

This chapter focuses on policies to attract human resources in science and technology. Many examples are given of policies being introduced in a wide range of countries, including policies for increasing the supply of highly skilled researchers, improving the attractiveness of the public sector, increasing the contribution of older researchers, repatriation and overseas recruitment of researchers, stimulating researcher links with industry and scaling up research infrastructures.

Introduction

The objective of this chapter is to present a short, policy-relevant analysis of the drivers of mobility of highly-skilled human resources, in particular international mobility, and examples of policy initiatives in OECD countries to attract and retain highly skilled workers. This information should help Scottish Enterprise and other development agencies in the design and execution of human resource-based innovation strategies.

The policy challenge

Scotland is at a cross-road. It has successfully attracted foreign direct investment, increased knowledge intensive exports in ICT, developed a growing biotechnology industry around high quality universities and increased exports in manufacturing as well as "niche" markets (*e.g.* in video games). Much of Scotland's success has drawn on fundamentals, first and foremost, a well educated and trained workforce and a small financial services industry that has successfully globalised itself and expanded into equity markets to become Europe's 6th largest market for investment management. But there are clouds in this picture. Scotland's Business Expenditure on R&D (BERD) is less than 0.5% of GDP or below half of the UK average. While business R&D jumped by 25 per cent in 2000-2001, most of the increase has been concentrated in branches of foreign firms. FDI which accounts for 14 per cent of output and 10 per cent of employment has declined recently, resulting in employment losses. The impact of FDI on supply-chains has been low as multinationals are plugged into global supply chains, making it difficult for small Scottish companies to access procurement markets. Scotland, while having a highly educated workforce, ranks low in the number of researchers and research personnel in the workforce. Researcher salaries in Scotland are below average salaries in Switzerland, Germany and the United States which makes attracting top talent difficult. The challenge facing policy makers at Scottish Enterprise and the Scottish Executive is how to stimulate and spread growth; how to move Scotland to the top quartile of Gross Domestic Product (GDP) per head among OECD countries.

Drivers of growth: The role of innovation and the highly skilled

The ability of Scotland to sustain durable growth rates in order to bring per capita income closer to that of the most advanced OECD countries cannot

 GLOBAL KNOWLEDGE FLOWS AND ECONOMIC DEVELOPMENT – ISBN 92-64-10766-5 – © OECD 2004

depend on one strategy however. The lessons of OECD countries over the past decade provide some insight into the question of what drives growth.

First, economic growth can be achieved by an increase in or improved use of labour, skills and capital as well as by increasing multi-factor productivity. OECD evidence shows that countries that improved growth in the 1990s benefited from the contribution of all of these factors. A large part of the growth in OECD countries has been due to improvements in labour utilisation – more people are working more productively. Another driver of growth has been improvements in the quality of labour. In many OECD countries, never before have young people been as educated as they are now. Of course R&D and technological progress also contribute to growth insofar as it improves multi-factor productivity growth. However countries that perform R&D and innovate are more likely to benefit from R&D done elsewhere. Furthermore, higher skill levels and use of information technologies are among two factors that enabled some countries to reap the growth benefits from investment in science and education (OECD, 2001). There is some evidence that sustained growth rates in productivity in the US over the 1990s have benefited from large inflows of foreign labour both unskilled and highly skilled.

In sum, the recent experience of OECD suggests there is "no silver bullet" for growth. Policy measures must be taken in a broad range of areas from higher education and skills, to R&D, and the promotion of ICT use in industry. In addition, while micro-policies for growth play an important part they must go hand in hand with sound macro-economic policies, policies which are set in London not in Edinburgh. But with devolution, Scotland does have some new cards to deal with. The *Science Strategy* and Learning and Skills initiatives are steps in the right direction.

The skills strategy

To improve innovation performance and to stimulate growth, investment in skills and in attracting and retaining talent in Scotland is one of the pillars of the Scottish Enterprise's strategy as articulated in the *Global Connections Strategy* and the *Smart Successful Scotland Strategy*. Scotland delivers 20 per cent of university graduates and 28 per cent of graduates in the life sciences but only represents 9 per cent of the UK population. There is also concern that many Scots that have emigrated to South East England or overseas are unaware of the opportunities in current day Scotland. Part of the plan focuses on enlarging the talent pool by both retaining existing Scots and attracting talent from overseas (although not necessarily returning Scottish migrants). The rationale behind this is recognition that highly skilled talent is an attractive force not only for FDI but also for R&D investments. Related to this strategy are efforts to link up Scotland's science base with the production system and to foster greater co-operation between the public research sector and industry. While Scottish Enterprise's Local Enterprise Companies (LECs) have long assisted existing firms in upgrading their human capital base and use of technology, in particular ICT, a new initiative, the "Intermediate Technology Institutes" (ITIs) aims to bridge the gap between invention and innovation. Rather than establishing "cathedrals in the desert", the ITIs will operate more as research granting agencies focused on specific clusters where strong interaction already exists between research, industry and universities.

Box 3.1. **Intermediate Technology Institutes**

Objective: the aim is to transform scientific discoveries into commercially viable products and in the process translate academic expertise into industrial strength and create thousands of high-end jobs.

Funding: £450 million to be distributed over a 10-year period.

Fields/Clusters: Energy research (Aberdeen), life sciences (Dundee) and communications technology and digital media (Glasgow).

Intellectual Property: ITIs will retain ownership of intellectual property (IP) created with funds.

Attracting talent from abroad

Across OECD countries migration of highly skilled personnel is increasing. The drivers of this increase are many but one important element is the preference given to skilled migrants in immigration policies. Indeed, the globalisation of higher education markets and demand for specialist talent combined with reforms to immigration systems that favour the highly skilled have all contributed to the global increase in international migration of students and highly skilled personnel. Like other parts of the world, the United Kingdom and Scotland have become more aware of the potential contributions of highly skilled talent to the economy.

But who are highly skilled we want to attract?

There are various forms of international mobility and various types of highly skilled migrants: students vs. workers, temporary vs. permanent workers; leading research scientists, IT specialists, research scientists, entrepreneurs. All of these migrants respond to different incentives and hence policies, like demand, will need to differentiate.

Drivers of international mobility

International mobility is a mechanism that helps supply and demand for labour to adjust on a global level. Information about opportunities and

 GLOBAL KNOWLEDGE FLOWS AND ECONOMIC DEVELOPMENT – ISBN 92-64-10766-5 –

conditions are quickly identified by workers, and potential migrants are identified by universities and firms which want to attract them. Moreover, with decreasing travel costs, it becomes easier for workers located abroad to keep ties in their country of origin, making migration a less costly and less irreversible decision than before. Like other categories of migrants, skilled people mostly move in response to economic opportunities abroad that are better than those available at home as well as in response to the migration policies in destination countries. Other factors, however, also play a role in the decision of the highly skilled to migrate and in their choice of destination and include intellectual pursuits be it education, research or language training. In the case of researchers and academics, the conditions in the host country regarding support for research and demand for R&D staff and professors can be an important determinant in the migration decision and destination. Among the entrepreneurially-minded, the climate for innovation generally and for business start-ups and self-employment in particular may play an important role in the decision of the highly skilled to move abroad. Another factor of mobility relates to the globalisation of firms that changes the geographical pattern of demand for specialist talent and feeds the transfer of skilled workers to other countries. For example, in the mid-nineties, intra-company transferees accounted for 5 to 10% of the total flows of skilled workers from Canada to the United States. As a region in the EU, Scotland benefits from access to the free movement of labour for EU countries. As an English-speaking region, Scotland also benefits from the demand for highly skilled students to improve language training. In addition, the language facilitates the integration of foreign students and workers.

Differentials in opportunities across countries contribute to explain the choice by certain individuals to migrate:

- Quality of the higher education system: the best students around the world may prefer to go to the best universities, wherever they are. And part of these students will stay in the receiving country after they have completed their studies.
- Job opportunities in certain industries, such as IT, attract foreign talent. There is an important role of institutions outside the labour market, such as venture capital, which help the expansion of high tech industries and in turn, demand for human resources in science and technology (HRST).
- Conversely, the lack of opportunities in certain sectors, (*e.g.* the public research sector in France, or lagged development of venture capital in Europe) may push people to move, such as young researchers with no job or would-be entrepreneurs with no access to capital.
- Quality of the working conditions, especially with regards to learning opportunities. Migrants tend to favour places where certain high tech

industries are already developed ("clustering": specialists go where other specialists are already located).

- Wage differentials have a clear impact, especially for flows between countries at a different stage of development (say, India and the US).
- Increasing international trade and FDI generate the need for multinational companies to allocate their staff at a worldwide level.

Evidence from a recent UK survey of non-EU migrants working in the UK on work permits showed that the dominant reasons for migrating to the UK were the career ambition of workers and opportunities to work in global centres of excellence as well as personal development. Five per cent of respondents in the biotech sector resided in Scotland in 2001 compared to 32 per cent in London and 26 per cent in the South East of England. In the health sector, 8 per cent of respondents resided in the London area while the share residing in Scotland was 13 per cent. In Finance the figures were 86 per cent in London and 2 per cent in Scotland (UK Home Office, 2002).

Economic impact of international mobility of highly-skilled workers

Foreign workers bring their skills and an increase in the quantity and diversity of the pool of skills available at the national level. They also help to keep wages moderate, reducing the cost of skilled labour for firms. This could also, however, be a negative effect for domestic workers in science and technology (S&T). Yet this is a static view which may miss dynamic effects. Because there are increasing returns to scale in knowledge industries, or agglomeration effects (spillovers), greater availability of S&T personnel may enhance the development of knowledge based activities, generating in turn more demand for S&T skills. For instance: more foreign entrepreneurs, can lead to more start-ups and enhance the demand for HRST, be it national or foreign. In other words, immigration of S&T staff increases the demand for all skilled labour, be it foreign or local.

Effects on higher education and research systems

For the sending countries, it is clear that absent the return of foreign students or workers, the investment in human capital is lost. The recipient country did not have to pay for the investment in education. For the host country, this is a short term advantage; in the long run this could lead to underinvestment in education among nationals. A strong out-migration of research staff may have severe impacts on the R&D and the wider innovation system. If emigrants return back, however, they bring back with them the knowledge and capacities they have acquired abroad. More than 10% of new recruits at France's National Research Centre (CNRS) came directly from a post-doc in the United States in 2000. Hence this "brain circulation" may dampen the negative effects, or even reverse them. Sending countries may be better integrated in knowledge markets.

By increasing the return on investment in human capital international mobility generates more attraction to education among young people. This may be in contrast with the political system's willingness to invest in education as some of this investment is lost.

Impact on innovation and entrepreneurship

In general, research shows there are several net positive effects for the main host countries, notably the stimulation of innovation capacity, an increase in the stock of available human capital and the international dissemination of knowledge. The contribution of foreign-born scientists to science is illustrated for instance by numbers of Nobel prizes awarded to US based researchers of European or Asian origin; for example, 32 % of US Nobel-prize winners in Chemistry between 1985 and 1999 were foreign-born. Skilled migrants are also a source of high tech entrepreneurship, in particular in geographical areas with strong concentration of higher education institutions and venture capital markets. It is estimated that a quarter of Silicon Valley firms in 1998 were headed by immigrants from China and India, collectively generating almost USD 17 billion in sales and 52 300 jobs. Often cited examples of successful firms that were in whole or in part created by immigrants include Inktomi, Ebay, 3Com, and Intel. It is clear however that the sustainability of such benefits for the recipient country requires that the immigration of foreign talent be accompanied by strengthened national investment in higher education, so that the pool of national talent is not depleted.

The mobility of skilled workers can also promote investment in training in sending countries and increase inflows of currency through remittances. Many of the benefits for sending countries, however, can only be realised in the longer term and require that countries invest in science and technology infrastructure and the development of the opportunities for teaching, research and entrepreneurship at home. The case of Chinese Taipei, as well as Korea and Ireland, suggests that when former skilled migrants come back to their origin country after a long stay abroad, their contribution to the expansion of a national high tech industry is considerable (OECD, 2001).

Immigration of HRST often correlates with inflows of financial capital. As these also tend to agglomerate around major knowledge institutions or local knowledge markets, the effects are very uneven within countries. Some of the effects may differ in OECD vs. non-OECD (less developed) countries.

Diaspora networks

Emigrants constitute a network abroad, which is helpful for firms which want to establish an affiliate or to export, which facilitates the circulation of knowledge from the host country to the country of origin. In both the US and the UK, there have been reports of successful Indian entrepreneurs establishing branches or even firms back home in India. But Diaspora networks also exist among like minded communities in developed countries. Examples include grass-roots initiatives of scientists and researchers in France and Switzerland to stay in touch with colleagues in the United States.

The importance of promoting domestic mobility

While international mobility can bring overall benefits to both sending and receiving countries, mobility within national labour markets, across regions etc can also stimulate a better match between supply and demand and contributes to increase the diffusion of scientific knowledge. Furthermore, there is some evidence that the greater mobility of workers correlates with multi-factor productivity growth (OECD, 2003). Mobility is a high priority in countries where there is a perception of low movement and where public employment systems are less flexible. Addressing mobility, however, is a question of both incentives and regulatory conditions. Competitive funding for research in the UK for example, indirectly fosters mobility of scientists at universities as "researchers follow the money". Flexible employment arrangements at "centres of excellence" are another way to foster researcher mobility. Fixed-term employment systems, while presenting other challenges, can also foster mobility.

In the context of devolution and greater autonomy, universities can also play a more active role in fostering movement between academia and the science base. In the Netherlands, encouraging mobility is a new objective of universities where, as a result of decentralisation, they have been given greater control of human resources management. The scope for fostering mobility differs between research institutions and universities. Support to mobility has traditionally been strong at the Dutch Foundation for Fundamental Research on Matter institutes and the TNO institutes. Mobility is also lower or higher depending on scientific discipline; data on Dutch mobility show that researcher mobility is lower in the humanities and the social sciences. As mobility generally concerns younger researchers, promoting mobility of older researchers is another challenge.

Policy initiatives from OECD regions and countries

In order to maintain current levels of research staff in the public sector and to attenuate the effects of an ageing scientific workforce witnessed in several OECD countries, a sufficient inflow of highly skilled researchers in key areas will

 GLOBAL KNOWLEDGE FLOWS AND ECONOMIC DEVELOPMENT – ISBN 92-64-10766-5 –

be required. Thus, attracting young scientists for a career in the public sector by improving the attractiveness of the public sector is a challenge to policy makers. Salary levels and the quality of the research infrastructure are important incentives for researchers to pursue a career in the public sector. In addition, the availability of PhD places and funding for doctorates have a significant influence on the decision of young researchers to enter a career in the public science and technology (S&T) sector. Initiatives for renewal of the public sector have been undertaken in many OECD countries: graduate education has been reformed, funding for PhD training has been increased and reforms in the overall structure of the public sector, *e.g.* changes in the tenure structure, have been undertaken.

Increasing the supply of highly skilled researchers

In order to increase the supply of young graduates in scientific areas, the Australian government has provided AUD 151 million over five years for an additional 2 000 university places each year with a priority in biotechnology and ICT to address shortages in these areas. In the UK, the Government and Research Councils have increased funding for PhD studentships and have committed to providing further resources for higher education to recruit and retain academic staff in science and engineering. The Dutch government has launched a special programme, the so-called "Renewal Impulse", which aims at retaining more young researchers in the public science system. In the period 2000-2010, 1 000 researchers will be selected for this programme. Furthermore, foreign students will be targeted for science-related careers in the Netherlands. In Germany, reforms aimed at shortening the doctoral programmes have been launched. Additional measures include a strengthening of the positions of junior staff in German universities and increased funding for research in high demand areas. In Sweden, during the "Promotion Reform" launched in 1999, 1 100 lecturers in higher education were promoted to the rank of professor. The Czech government in their National Research and Development Policy committed to improving the material situation of young R&D workers and to increasing the funding for talented young researchers. In order to attract graduates into public research the Hungarian government increased the salaries for public sector researchers, especially for young graduates, in 2001.

Improving the attractiveness of the public sector

Salaries and research conditions are key incentives for young researchers to pursue employment in the public sector. They are also important in preventing an internal as well as an international "brain drain". The quality of graduate education is seen as one of the reasons for the attractiveness of the US system: the existence of a dense network of high-quality research facilities

Box 3.2. **The Canadian strategy for enhancing human capital**

Strategy

The Government of Canada has made the development of human capital one the main pillars of its innovation strategy. It has set clear goals, targets and federal priorities to develop, attract and retain the highly qualified people.

Goals: Develop the most skilled and talented labour force in the world. Ensure that Canada receives the skilled immigrants it needs and helps immigrants to achieve their full potential in the Canadian labour market and society.

Targets: Increase the admission of Master's and PhD students at Canadian universities by an average of 5 per cent per year through 2010. By 2004, significantly improve Canada's performance in the recruitment of foreign talent, including foreign students, by means of both the permanent immigrant and the temporary foreign workers programmes. Another target is the increase by 1 million of the number of adults undergoing life-long learning.

How to implement the strategy:

Providing financial incentives to students registered in graduate studies programmes, and doubling the number of Master's and Doctoral fellowships and scholarships awarded by the federal granting councils. Creating a world-class scholarship program of the same prestige and scope as the Rhodes Scholarship.

Supporting and facilitating a co-ordinated international student recruitment strategy led by Canadian universities. Implementing changes to immigration policies and procedures to facilitate the retention of international students. Establishing a co-operative research programme to support graduate and post-graduate students and, in special circumstances, undergraduates, wishing to combine formal academic training with extensive applied research experience in a work setting.

allows young researchers to pursue high-quality research close to their field of degree. The introduction of Integrative Graduate Education and Research Traineeship (IGERT) programmes offering stipend support to graduate students to engage in research in emerging multidisciplinary areas in science and engineering have contributed to this development.In Hungary, a closer co-operation with industry, *e.g.* the establishment of R&D labs in universities, aims at improving the research environment of young researchers.

Increasing the contribution of older researchers

The experience of older researchers is important for transferring knowledge and know-how to new researchers. The challenge for policy makers is to attract a sufficient number of young researchers while offering flexible work arrangements for older researchers. Reforms aimed at lengthening work life for older researchers (*i.e.* in the 55 to 64 age group) might lead to a further increase in the share of scientists in this age group, but would also contribute to an increase in the total number of researchers in the public sector. In most OECD countries strong incentives to early retirement exist. An increased flexibility with regard to the work time or retirement schemes of older researchers might have positive effects on both the age distribution of the overall scientific workforce in the public sector and the knowledge diffusion between generations. Part-time work of older scientists might lead to better career prospects for young scientists. In areas where a growing demand for scientists is forecasted and the supply of sufficient numbers of new researchers is not yet assured, such flexible arrangements might contribute to a smooth adjustment of research systems.

Repatriation schemes – Mobility with strings attached

Promoting mobility without endangering the national scientific base (*i.e.* a brain drain) is a key goal of policy makers. Support for international mobility has traditionally been focused on supporting the temporary outward mobility of post-docs and researchers, but there is a new emphasis on attracting foreign researchers in order to increase supply and access specialist skills. Most OECD countries maintain schemes to help students and post-doctorates to study/work abroad on a temporary basis. At the EU level, the European Commission schemes such as ERASMUS (for students) and Marie Curie Fellowships for researchers aim to increase intra-European mobility. As part of the ongoing reform of Norwegian higher education, students are encouraged to spend at least one term of their course studies at a foreign institution. The home institution facilitates such mobility and receives financial reward for the international exchange of students.

Large immigration countries such as Australia, Canada and the United States have long relied on foreign students and researchers to meet demand in the national science system and to supplement innovative capacity. Especially in the 1990s, US academic institutions and the US science and engineering workforce as a whole have relied on foreign-born (often US-educated) persons; in some engineering and computer science fields, they exceed one-third of the total. In addition, a number of visa classes (in addition to the H1B temporary visa for highly skilled professionals) facilitate the temporary move to the US of highly educated personnel, and there is discussion at the US Congress about liberalising spousal work permit rules. In the past two years, however, Germany and the United Kingdom, and to a lesser extent France, have also made the attraction of top foreign students and researchers a priority (see Box 3.3).

Box 3.3. **Science and technology policies to retain and attract scientific talent**

Attracting foreign and expatriate talent. The UK government, together with the Wolfson Foundation, is funding a Research Merit Award scheme, run by the Royal Society and worth £20 million over five years. This offers institutions additional funds to increase the salaries of researchers whom they wish to retain or recruit from industry or overseas. In Germany, the Humboldt Foundation and the German Federal Ministry for Education sponsor a EUR 22 million Research Award the "Sofja Kovalevskaja-Preis" to help young scientists from overseas, as well as expatriate German scientists, carry out research in Germany for a period of 3 years. A single award can be as much as EUR 1.2 million. France has long supported the temporary stay of foreign researchers but a new initiative launched in 1999 aims to attract some 200 young researchers each year, in particular from emerging economies such as Brazil, China, Mexico, and South Africa.

Modernising the immigration system. The Government of Canada is updating immigration rules to maintain higher immigration levels and increase the number of highly skilled workers. In addition, it plans to improve the efficiency of the migration process and to enhance the attractiveness of Canada as a destination. Specific initiatives include using a redesigned temporary foreign worker programme and expanding provincial nominee agreements to facilitate the entry of highly skilled workers, and to ensure that the benefits of immigration are more evenly distributed across the country.

Providing tax incentives to encourage recruitment of foreign personnel. Denmark, the Netherlands and Belgium have passed laws to alleviate the tax burden on foreign experts and highly skilled workers. In Quebec, the government is offering five-year income tax holidays (credits) to attract foreign academics in IT, engineering, health science and finance to take employment in the province's universities. In 2001, Sweden adopted similar policies for highly skilled workers who live in Sweden for less than five years.

Repatriation schemes for post-docs and scientists. The Academy of Finland has a programme to ease the return to Finland of Finnish researchers, who have been abroad for a length of time. In Austria, the Schrödinger-scholarships help returning Austrians integrate into scientific institutions. Germany's Ministry for Research and Education (BMBF) has also launched a new programme in 2001 to attract the return migration of German researchers overseas. In support of the repatriation of Canadian postdoctoral researchers, the Canadian Institute for Health Research offers a supplementary year of funding to Canadians and permanent residents who are recipients of either the Japan Society for the Promotion of Science (JSPS) Postdoctoral Fellowships for Foreign Researchers or Welcome Trust/CIHR Postdoctoral Fellowships. In order to be

Box 3.3. **Science and technology policies to retain and attract scientific talent** *(cont.)*

eligible for the "Canada Year" funding, training must take place in a Canadian laboratory. Italy has recently introduced the "Reverse Brain Drain Project", which is aimed both at attracting foreign professors and scientists and at facilitating the repatriation of Italian scholars abroad. In 2002, the Italian government funded additional positions with EUR 20 million. Over 100 foreign scholars have been employed in Italian universities, most of them in the fields of mathematics and physics (51%) and engineering. Also, 63 Italian scholars benefited from the project. The Italian government will continue this project in 2003.

Leveraging immigrant and Diaspora networks. Such networks do not only exist among emigrants from developing countries; Swiss scientists in the US have created an Internet network and directory (*Swiss-list.com*) to link Swiss scientists and post-doctorates working in the US to colleagues in Switzerland. The French foreign ministry sponsors meetings between French post-doctorates working in US research institutions and French companies.

Source: OECD (2002), *Ad hoc* Group on Steering and Funding of Research Institutions Questionnaire Results; *The International Mobility of the Highly Skilled*; 2002.

Stimulating inter-sectoral mobility

Linkages between the research sectors exist at both formal and informal levels and may promote inter-sectoral mobility. Joint location of university and public institute sectors is common in some countries. In France, many of the laboratories of the "Centre National de Recherche Scientifique" (CNRS) are located on university campuses and consequently, staff from both sectors may exchange experiences in the workplace. In Germany, the "Max Planck Gesellschaft" (MPG) has followed a policy of establishing an institute in proximity to a university with a focus on a similar research area, thus enabling researchers to undertake joint work, but from different bases. Recently, however, jointly located teams have been established within universities. Also, directors of institutes with major networks often hold an additional post at local universities. Research training is also a key area of collaboration between the sectors. Although only universities have the authority to award research degrees, many doctoral students who work in public research institutes in Norway use institute staff as supervisors for their theses.

Across OECD countries universities and education systems are gaining in autonomy *vis-à-vis* education and research ministries. In federal countries,

Box 3.4. Promotion schemes for researcher mobility and co-operation with industry

Austria maintains mobility promotion schemes such as "Scientists for the economy" and the mobility of junior researchers is promoted through the Industrial Promotion Fund. The "Social Science Fund (FWF)" has envisaged the creation of "Graduate Program (WK)" centres for the education of highly skilled young scientists. They will be established in scientific areas where the productivity in Austria is exceptionally high.

Australia's Linkage-Projects scheme, which is administered by the Australian Research Council, supports collaborative research projects between higher education researchers and industry. Under this scheme, support can be provided for: Australian Postdoctoral Fellowships Industry (APDI) for researchers with less than three years' postdoctoral experience; and Australian Postgraduate Awards Industry (APAI) for postgraduate research students studying towards a Masters or PhD award.

Canada: The Natural Science and Engineering Research Council sponsors postgraduate training in industry through various schemes including scholarships for training masters and doctoral students in industry and fellowships for the hiring of recent PhD graduates by firms.

France: The Ministry of Research fosters PhD training in a research company by subsidising up to half of the corresponding salary costs to the firm. Subsidies for post-doctoral positions in SMEs are available to young PhDs without industry experience.

Hungary: The knowledge flow and mobility between research institutes, higher education and industry will be promoted by the foundation of CRCs.

Japan's latest Basic Science and Technology Promotion Plan outlines a series of regulatory reforms to the labour market for public sector research, aimed at improving mobility between the public and private research sectors. The *Centres for Co-operative Research* in 56 national universities carry out joint industry-public research as well technical training of researchers from private companies. A main goal is to create critical mass by channelling individual researcher collaboration into institutional level linkages.

Korea: The Korean Institute of Science and Technology (KIST) has promotional schemes to grant temporary leave to researchers to undertake entrepreneurial activities.

The Netherlands' KIM scheme that promotes the movement of S&T personnel to SMEs has shown success. Furthermore, under the WBSO (Act to promote R&D) small firms are allowed a tax deduction for the labour costs of R&D staff.

GLOBAL KNOWLEDGE FLOWS AND ECONOMIC DEVELOPMENT – ISBN 92-64-10766-5 – © OECD 2004

Box 3.4. **Promotion schemes for researcher mobility and co-operation with industry** *(cont.)*

Norway has set up special programmes to stimulate mobility from universities/research institutes to the private sector and to make industry-relevant research more attractive, such as the FORNY programme that is entering its third phase.

Portugal: The Ministry of Science and Technology runs a programme to help the placement of new PhDs in firms through the subsidisation of salaries for 2 years.

Sweden: The NUTEK competence centres at universities promote collaboration between public researchers and those in firms which may help break down non-regulatory barriers to mobility.

United Kingdom: The Faraday Programme promotes a continuous flow of industrial technology and skilled people between industry, the universities and intermediate research institutes. In 1999, it was expanded with a focus on entrepreneurial activities and research commercialisation. In addition, the long established Teaching Company Scheme finances an associate to work on project in a semi-academic or company environment for two years. Both schemes are operational in Scotland.

United States: The Grant Opportunities For Academic Liaison with Industry (GOALI) initiative of the National Science Foundation (NSF) funds: 1) faculty, postdoctoral fellows and students to conduct research and gain experience with production processes in an industrial setting, 2) industrial scientists and engineers to bring industry's perspective and integrative skills to academia and 3) interdisciplinary university-industry teams to conduct long-term projects. There are no requirements for matching funds from firms for GOALI projects performed in universities. University-industry IPR agreements must be made up front and submitted for funding consideration.

Source: OECD (2002), *Ad hoc* Group on Steering and Funding of Research Institutions Questionnaire Results, *Benchmarking Industry-Science Relationships*, forthcoming 2002.

such as Canada, Germany and the United States, the trend to autonomy is a natural consequence of decentralisation in policy making and funding. Although the UK is not a federal country, under devolution, Scotland has obtained greater autonomy in the spending of higher education institutions and in education and research policy. More autonomy grants universities and public research organisations more freedom with regard to human resources management including the hiring and setting of salaries. This freedom is relative however and is to some extent limited by the amount of core funds and employment agreements with the social partners. Shifts in funding also

Box 3.5. **Improving the attractiveness of the public sector**

Raising salaries and funding: The UK Government plans to increase the salaries of post-doctorates by 25% and increase funding for the hiring of university professors. The Czech Republic has implemented schemes to provide additional financial support to young R&D workers up to 35 years of age. The European Commission has doubled the amount of funding devoted to human resources in the Sixth Research Framework Programme to EUR 1.8 billion in order to improve the attractiveness of the European Research Area. The Backing Australia's Ability initiatives include establishing prestigious Federation Fellowships worth AUD 225 000 a year each. These are aimed at attracting and retaining leading researchers in key positions and up to 125 Federation Fellowships will be awarded with total funding of AUD 112.3 million over the next five years from 2002 to 2006. The Prime Minister announced the first fifteen Federation Fellowships on 25 September 2001. In addition, the number of Australian Postdoctoral Fellowships will be doubled from 55 to 110 and remuneration of these positions will be improved, with total funding of AUD 50.1 million over five years from 2002 to 2006.

Employment reforms and post creations: Germany is launching the development of Junior Professorships, temporary posts to attract young researchers to university employment in some 30 universities. These junior professors will be tied to research departments rather than to professors, which is currently the case for new academics. In 2001, the BMBF is providing EUR 6.1 million. Junior Professors are granted 3-year employment contracts, renewable once. In Austria, a major reform has taken place in the employment of the university system. As of January 2004, new university staff will not have civil servant status and employment contracts will be limited (four – six years) after which scientists/researchers will have to apply for new contracts, depending on the number of available posts. Tenure will only be granted to full professors. Currently 21-23% of total university staff are tenured professors. Norway aims at increasing the number of doctorates by 60% by 2007 in order to secure recruitment to research in academia and industry, international recruitment and the recruitment of women. In France, some 700 teaching-researcher posts were created between 1997 and 2001 to strengthen the public research sector and attract post-doctorates from overseas. The Netherlands Ministry of Education, Culture and Science and the Ministry of Agriculture, Nature Management and Fisheries, now together with the universities have launched the Renewal Impulse scheme to retain bright young researchers within the public science system. The programme focuses on three stages of the scientific career up to professor: young post-docs, experienced post-docs and top talent. In the first round (2000) 43 candidates were assigned. The impulse aims to select over 1 000 researchers in the period 2000-2010.

Source: OECD (2002), *Ad hoc* Group on Steering and Funding of Research Institutions Questionnaire Results.

affect employment structure and priorities. The general increase in the reliance of institutions on external (non-intramural) research has increased the number of researchers whose funding depends on external funds.

Box 3.6. **Attracting research and talent to Ireland**

Two government-administered schemes will distribute about 1.2 billion euros (US$1 billion) by 2007. The Programme for Research in Third Level Institutions (PRTLI), administered by the Higher Education Authority, was started in 1999 and has so far awarded over 600 million euros. A second scheme, Science Foundation Ireland (SFI), was launched in 2000 by the Department of Enterprise, Trade and Employment, and will give out 635 million euros in grants by 2007. The SFI grants will be made on merit review by top scientists and will target two fields: biotechnology and ICT.

The schemes aim to attract researchers of any nationality, including Irish scientists currently working abroad. Employment creation is one of the underlying goals of the programmes. PRTLI funding is also expected to double the number of post-doctorate positions over the next three years.

Increasing the number of academic staff through the SFI grants presents additional "challenges" as international candidates need to be sure that the investment will be sustained – especially if they would be moving their entire lab to Ireland.

The newly established Programme for Human Genomics, funded with EUR 44.8 million of PRTLI grant funding will require 40 graduate students, 40 postdocs and 20 academic staff.

Building infrastructure and scaling up

It is expected that PRTLI-funded centres will become self-sufficient, relying on external grants. Attracting grants will however require Irish universities to reach critical size as research institutions. The funds and a shift towards more R&D appear to be helping attract pharmaceutical companies. The firm Wyeth is investing USD 1 billion to build a biopharmaceutical campus outside Dublin. The firm will focus on drug manufacture as well as R&D.

Centres of excellence also help attract funds and people. Recent examples include the National Institute for Cellular Biotechnology at Dublin City University which was established with PRTLI funding (EUR 34.3 million). Building the infrastructure increases demand for new faculty, student and post-doctorates. Changing curriculums at higher education institutions is also necessary to improve the environment for R&D.

Sources: Science Foundation Ireland, *Naturejobs.com*.

Initiatives in non-OECD countries

Singapore

As a small country with little natural resources and a population of around 3 million people, Singapore is dependent on brain power, including foreign talent. Government statistics estimate that of the 4 000 research scientists, some 40% are foreign. Like other emerging regions, Singapore has carved out a place in the knowledge economy by supplying electronics and ICT related goods to firms world-wide. Anticipating the growing competition in ICT industries, the government has had the foresight to invest in other knowledge-intensive industries, in particular biotechnology. This has attracted R&D investment from foreign pharmaceutical companies such as GlaxoWellcome (UK) and Rhone Poulenc (France).

Building on national competences

To attract foreign R&D, the government, through the National Science and Technology Board, has focused its efforts on improving and expanding investment in higher education and research facilities as well as creating centres of excellence. A case in point is the Institute of Molecular and Cell Biology (IMCB) which was established in 1987. The 250 staff centre is highly competitive at international level and a magnet for researchers worldwide in the areas of functional genomics, immunology, virology and drug discovery.

Like Scotland, Singapore is trying to develop a critical mass of skills in targeted areas. New centres include the Institute of Molecular Agro biology (1995); the Bioinformatics Center (1996); the Institute of Materials Research and Engineering (1997); and the Cancer Therapeutics Research Group (2000) and the Singapore Synchrotron Light Source (2001). Singapore's National University is planning to establish a tropical marine science institute, capitalizing on its biodiversity. The institute will conduct research into marine biotechnology, oceanography, offshore engineering and tropical fish farming.

Recruiting researchers from overseas

To staff these institutes, the government is recruiting top talent from abroad. It therefore putting more money into fellowships for new postdoctorates and post-masters. The country is targeting young graduates in mainland China and hopes that once they have completed their two-year fellowships that they will move onto industry.

Like Scotland's Talent Scotland, Singapore has established an internet based job matching service that links up foreign faculty with firms and institutes in Singapore. Another *foreign researchers' programme* assists companies in the recruitment of experienced foreign researchers, providing

50 per cent funding of recruitment by helping with relocation costs, salaries and housing allowances for up to two years. The *Temasek Professorships* target individuals in areas in which Singapore wants to develop, and hire them to head or start laboratories. These areas include biological sciences, mathematics, semiconductors and data storage. Singapore plans to recruit between 20 and 30 *Temasek* professors for between 3 and 5 years and has allocated a budget of US$89 million. Recipients of the professorships, selected by a steering committee, will be required to spend only 50 per cent of their time in Singapore.

Policy implications

The above discussion has highlighted the tremendous amount of policy attention on human resources. Across OECD countries, regions and beyond, the issue of a well trained and highly skilled workforce is taking centre stage. On the supply-side, countries from Canada to the Netherlands are increasing the supply of S&T graduates and young PhDs. Countries are also encouraging young people to stay in science by providing post-doctorate research fellowships. Scotland already produces a large number and share of S&T graduates relative to its population. Where more action is needed is in encouraging young people to work in research. But first, policies must deal with the demand-side. Initiatives in Germany, the Netherlands, Ireland and Singapore to support young researchers in public research and make research more attractive are instructive. Like Irish universities, a challenge for Scotland is to increase the size of key research departments by hiring top faculty from within and outside Scotland. The ITIs are a good way to increase funding for research but they may also emulate efforts in Singapore and Ireland to attract the "magnets" of foreign talent. If you rake in a "big" fish you might get an entire school or rather research team with it. The ITIs should also be leveraged to help universities attract contract research from firms. This will stimulate greater cross-fertilisation between firms and universities and introduce young students to the industrial research environment.

Attracting talent from other regions or from abroad presents another set of challenges. The UK already has a high share of foreign PhDs relative to its population. Getting some of those talented graduates to stay in the UK and in Scotland should be an integral part of Scottish Enterprise's strategy. Stay rates of Chinese students in science and engineering in the UK are much lower than stay rates in the US (NSF, 2002). Evidence from the US shows that some 25% of foreign highly skilled workers on the H1B visas were already in the country as students before they changed immigration status. Recent immigration reforms in London that allow students to stay in the country and enter employment will help. While many OECD countries have specific programmes to attract highly skilled workers, it is important to note that although such highly skilled people may be versed in the language of the host country, their cultural adaptation and integration in the host region takes time and investment both from the side of the migrant as well as the hiring firm or university. In Finland, for example, Nokia invests in the cultural adaptation of foreign information technology workers as a way to improve productivity but especially the *retention* of workers.

In conclusion, Scotland's strength in human resources has served it well and continues to be a solid basis on which to grow its economy and raise levels of per capita GDP. But the knowledge economy requires continual investment in human capital and the renewal of a skilled workforce. Scotland now faces competition from other regions and countries for talent, both for its own talent resident abroad as well as for specialists and top researchers. The following policy recommendations could apply to Scotland, just as they could to other regions and countries. While a comprehensive approach is necessary, the challenge for Scottish Enterprise is to focus on the "weakest link" in its innovation systems, in particular as regards the human resources and skills base. Based on the OECD review panel visit, this weak link appears to be at the intersection of the incentives and opportunities for young people to work in research in Scotland. Initiatives like Global Scot and Talent Scotland are helping tackle part of the weakness in the link but more resources for R&D will be needed to create the critical mass to grow Scotland's centres of excellence into hubs and clusters of global innovation.

Create conditions at home for attracting and retaining talent

- Maintain support for higher education in science and technology.
- Foster framework conditions for growth, *e.g.* that favour entrepreneurship.
- Increase investment in public R&D, especially in the "R".
- Stimulate business R&D and greater interaction with the public research base.
- Promote co-operation between industry and universities in training young researchers.
- Provide incentives for Scottish emigrants to return back home with their newly acquired competences.
- Leverage diaspora networks (*e.g.* Global Scot).

References

Barro R. J. (2001), "Education and Economic Growth" published in Helliwell, The Contribution of Human and Social Capital to Sustained Economic Growth and Well-being. Proceedings of the HRDC/OECD Symposium on the Effects of Human and Social Capital.

Cervantes, M. and Guellec D. (2001), The Brain Drain: Old Myths, New Realities. *OECD Observer* March 2001.

Helliwell (2000), The Contribution of Human and Social Capital to Sustained Economic Growth and Well-being. Proceedings of the HRDC/OECD Symposium on the Effects of Human and Social Capital. Quebec, March.

Hood, Peat, Peters and Young (2002), *Scotland in a Global Economy: The 2020 Vision*. Palgrave Macmillan. Hampshire, United Kingdom and New York, United States.

National Science Foundation (NSF) (2002), Science and Engineering Indicators 2002.

Scottish Executive (2002), A Smart Successful Scotland.

Scottish Executive (2002), Scotland: A Global Connections Strategy.

OECD (1998), *Mobilising Human Resources for Innovation*, OECD: Paris.

OECD (2001a), *The New Economy: Beyond the Hype*, OECD: Paris.

OECD (2001b), Drivers of Growth: Information Technology, Innovation and Entrepreneurship: Special Edition of the *Science, Technology and Industry Outlook* 2001.

OECD (2001c), *International Mobility of the Highly Skilled*, OECD: Paris.

OECD (2003), The Management of Human Resources. Report of the Sub-Group of the CSTP *Ad Hoc* Working Group on the Steering and Funding of Research Institutions [DSTI/STP(2003)9ANNEX4].

Office of Technology Assessment (1987), Preparing for Science and Engineering Careers: Field Level Profiles, published by the Science, Education, and Transportation Program of the Office of Technology Assessment, US. Congress.

United Kingdom HM Treasury (2002), SET for Success: The supply of people with science, technology, engineering and mathematics skills. The report of Sir Gareth Robert's Review. April.

UK Home Office (2002), Knowledge Migrants: The motivations and experiences of professionals in the UK on work permits.

ISBN 92-64-10766-5
Global Knowledge Flows and Economic Development

Chapter 4

Integrating global knowledge flows for generative growth in Scotland: Life sciences as a knowledge economy exemplar

by

Philip Cooke (Cardiff University, United Kingdom)

The stimulation of entrepreneurship in knowledge-intensive industries is the focus of this chapter. Intermediary technology transfer agencies are considered critical to identifying and helping fund exploitable research and securing its commercialisation in new firms. Strategies to build the life sciences sector are highlighted as an example of how to meet the challenges of the knowledge economy. Recommended actions including increasing research funding and university training in the sector, raising awareness of existing talent amongst potential inward investors, seeking to commercialise innovation in new firms, encouraging networking and knowledge transfer in the sector and helping to access key knowledge from overseas.

Introduction

"Life Sciences" are a significant future sector for Scotland's economic development. Public research funding is running at £1 billion per year in the UK, USD 27 billion in the USA, and EUR 10 billion in the EU. Over 250 biopharmaceuticals products are in clinical trials worldwide and the global market is over US$100 billion. The field encompasses companies in biotechnology, pharmaceuticals, biomedical technologies, life systems technologies, nutraceuticals, food processing, environmental technologies, biomedical devices, and organisations and institutions that devote the majority of their efforts to the various stages of research, development, technology transfer and commercialisation. More generally "Life Sciences" may be defined as all sciences that deal with "organisms", like plants, animals and human beings. In this discussion the focus will be on the most significant current applications for Scotland, which relate to healthcare. The OECD mission to investigate the development and implementation of policies to enhance international knowledge flows represented an opportunity to observe innovative actions taken to re-position a national-regional economy by its Parliament (via the Scottish Executive) and principal economic development agency (Scottish Enterprise) faced with three fundamental challenges. The first of these is *Globalisation*, meaning the rise of extensive and intensive economic forces spread more widely and deeply creating greater integration and interdependence in the world economy. This is orchestrated by multilateral regulatory and advisory bodies such as the World Trade Organisation, United Nations, OECD, European Union, etc., to create liberal conditions for promoting global free trade, and multinational, or transnational corporations (TNCs) that are the agents of and significant beneficiaries of the achievement of trade liberalisation. In simplified terms, the form taken by *globalisation* can be stylised as the integration of global value chains and regionalised or localised industry clusters (UNIDO, 2002).

Hitherto, economies like Scotland have benefited from their capability in attracting the production end of global value chains, particularly in Information and Communication Technologies (ICT), to agglomerate in regional-local settings (the so-called "Silicon Glen"). However, the current phase of globalisation makes continuation of strategies in this way problematic. This is due to the propensity of TNCs to seek cheaper production

 GLOBAL KNOWLEDGE FLOWS AND ECONOMIC DEVELOPMENT – ISBN 92-64-10766-5 – © OECD 2004

bases elsewhere, in Central and Eastern Europe, North Africa, Latin America and Asia, particularly China (Morrison, 2002).

This is addressed in Scottish Enterprise's "*Global Connections*"strategy.

The second challenge facing peripheral regional economies that restructured away from dependence on heavy industry by adapting their "employee culture" to Foreign Direct Investment (FDI) is that, with the decline in growth opportunities for inward investment, *Entrepreneurship* has become something of a policy mantra everywhere.

This is *generative* as distinct from *redistributive* growth. It is cognate with the neo-classical notion of "endogenous growth" (Romer, 1990) but rests on a more socialised view of the individual, the firm and the economy than neo-classical theorems allow, as will be clear from what follows below. Entrepreneurship means capability to mobilise skills and resources to create a sustainable business. This is a talent that the entrepreneurship literature suggests is possessed by perhaps 10% on average of the working population, although it accounts for over 50% of employment and GDP, and some 97% of firm numbers in mature economies like Germany or the UK (OECD, 1994). Since the 1970s "mobilisation of indigenous potential" has been an injunction of economic organisations like the EU and OECD, early in recognising the beginnings of "downsizing" by multinationals and other large manufacturing firms in the face of, first, Japanese then other Newly Industrialising Countries (NICs) like South Korea and the other "Asian Tigers", Mexico, and Brazil. Contemporaneously, through the 1980s, many countries, led by the UK, sought to reduce investment and employment in the public sector and privatise the parts that had market potential. Competitive pressures have seen major reductions in employment in such industries. Hence, only the small and medium-sized enterprise (SME) sector seemed likely to offer opportunities for employment growth. This was because the elaboration of value chains by large firms responding to heightened competition by cost reduction through externalisation of inputs, and the growth of "compulsory competitive tendering" by bureaucracies obviously favoured SMEs. To which could be added the opportunities for innovation-based entrepreneurship arising from scientific or creative industry discoveries that again favoured SME development. However, to reach the average 10%, let alone grow shares of entrepreneurs above it is extremely difficult, and it is particularly so in "employee culture" economies, where generally speaking shares are nearer 6% or less to begin with. This is because entrepreneurship both thrives and gravitates towards areas with an "entrepreneurial social environment" (Flora, Sharp and Flora, 1997), and places with "economic business environments" (Porter, Sacks and Warner, 2000). This means marketised services in finance, law, management accountancy, consultancy, contracting etc. by means of which individuals may pay to have "the pain taken away" in establishing and managing sustainable businesses (see, *e.g.* Kenney, 2000; Lee, 2000). These problems, arising from market failure occasioned by employee "path dependence" (Arthur, 1994) are recognised in Scottish Enterprise's "*Growing Business*" strategy.

Finally, the Scottish economy like every other is challenged by the rise of the *Knowledge Economy* conditioned by science-based industry (and to some extent creative or culture-based industry, notably media and new media; Cooke, 2002). It is widely considered that modern life sciences is easily the most science-driven of "knowledge economy" industries, and for that reason extra attention is devoted to it below. This is also because it is a key in both the Scottish Executive's "*Science Strategy for Scotland*"and Scottish Enterprise's ambitions to facilitate growth in the emergent Scottish biotechnology sector by stimulating industry "clustering". But its influence is widely evident, having major growth implications in healthcare (held to account for 25% of UK GDP in 1999 (DTI, 1999), and rising following the UK Treasury's commitment of a further £40 billion over five years from 2003 for the National Health Service), as well as agro-food and environment, not to mention ICT and nanotechnologies. This means it will exert significant weight in business opportunity of many kinds in the foreseeable future, and inevitably influence other business activity as a successful role model. This is due to the symbiotic relationship of dedicated biotechnology firms (DBFs) with both well-funded drug companies ("big pharma") (global 2002 R&D budget $28 billion) and public research funds (*e.g.* 2003 US Federal Health R&D budget appropriation of $27.3 billion). This means relatively few DBFs "go out of business" once established, especially in therapeutics, even though few make what conventional accounting calls "profits". To build such industry requires a major expansion in the pool of scientific, innovative, entrepreneurial and creative "talent". This is recognised in the "*Learning and Skills*" strategy of Scottish Enterprise, though it is set far wider than life sciences, as it needs to be.

Scotland's strategic framework

Clearly, thus far, Scotland may be thought fortunate in having in place strategies to meet three of the key forces for economic change currently being faced worldwide. However, because of their ubiquity these forces are being met in various ways appropriate to national or regional conditions throughout the world. It is worth noting, too, that nowadays, leading-edge policy thinking is transmitted at nanosecond velocity not only among OECD member countries but also from them to developing countries, through the activities of private consultants and multilateral agencies like UNIDO. Hence, to return briefly to the challenge posed by China, it is instructive to note that the speed of its development, measured by its competitive industrial performance 1985-1998, is officially estimated as treble that of its nearest OECD competitor, Ireland, and ahead (marginally) of its nearest Asian competitor, The Philippines (UNIDO,

2002). The Chinese strategy for translocation of global ICT production *and* R&D into the Beijing, Huangdong (Shanghai) and Guangzhou (Shenzen-Guangdong) regions has borne fruit, significantly on the back of an investment in engineering talent that, at 37% of graduate output, is more than six times that of the US at 6% (Morrison, 2002). China also has a developing Life Sciences strategy in place. Currently it vies with India as the leading global location for the important, if medium value chain activity of *biomanufacturing*, for which there is currently a major global capacity shortage. As elsewhere, the ICT regions are also slated to host biotechnology R&D and exploitation, not least drawing on the country's long and respected tradition in natural healing therapeutics.

But we need not look as far as China to observe ICT and Life Sciences led strategies evolving into leadership of future economic development portfolios in Finland, Sweden, Denmark, Ireland, Netherlands, France, Germany, UK, Israel, Canada, Japan and, of course, the USA. A similar, more variegated picture could be presented at regional level for many of the aforementioned countries. Some, notably the US, possibly Canada and other parts of the UK, are more advanced than Scotland in the healthcare application of Life Sciences discoveries, though not in some basic science and its applications, notably in transgenics (PPL Therapeutics) and anticancer (*e.g.* Cyclacel's World Economic Forum "pioneer" citation in 2002). Past recognition of the unsustainability of an FDI strategy of the kind practised hitherto, and the difficulty in inducing an indigenous value chain, as the main means of building an ICT industry led ultimately to the Alba Centre, a facility aimed at attacking the upper reaches of the value chain by training and spinout in advanced software. With Alba as something of a prototype, new Intermediate Technology Institutes (ITI) are now planned for Bioscience, ICT and Energy exploitation and commercialisation (see Appendix 1). Thus Scotland, having cast a wide net, has learned, notably, it may be inferred, from Taiwan, about the importance for commercialisation of innovation through an intermediary entity like Taiwan's Industrial Technology Research Institute (ITRI) founded in 1973.

From the perspective of managing international knowledge flows, "catch-up" was dependent less on basic research quality from indigenous Higher Education Institutes, than on forging alliances with TNCs like IBM and Motorola, then transferring technological knowledge to receptive SMEs for commercialisation. Taiwan's current dominance in mobile PCs rests on the work of such public-private consortia that rushed product to world markets in 1991. Its strong performance in data switches, crucial to PC networks, also arose from ITRI-SME consortia to match the Ethernet standard in 1993. The IBM PowerPC microprocessor of 1995 was cloned simultaneously through ITRI's alliance with originating designers IBM and Motorola. Subsequent successes arose via the partnership mechanism in digital communication and multimedia areas. More recently, the Taiwan Engine Company was established by welding three separate SMEs together in alliance with ITRI to supply 1.2 litre 4-valve engines to the burgeoning Chinese automotive market (Mathews and Cho, 2000).

Adapted to the Scottish context, where the prospects for commercialising original, as distinct from cloned, knowledge are higher than they were in Taiwan, yet the university originators are, as elsewhere, inexpert at generic, swift-growth spinout incubation, and the innovative, technology-intensive SME sector is rather thin, ITIs have a logic to them, given the knowledge transfer and translation imperatives of turning exploration knowledge (basic research) into commercially exploited innovations. Where might such an institution meet barriers? Three areas are immediately evident. First, ITRI and comparable bodies in other Asian countries like Korea's Regional Research Centres (RRCs) or Singapore's Information Technology Institute, once the research arm of the National Computer Board (now Infocomm Development Authority – IDA), or government-owned Singapore Technologies, through its subsidiary Chartered Semiconductor Manufacturing (CSM), make deals and alliances directly with TNCs. In the CSM case the partner in a joint wafer fabrication plant is Lucent Technologies; for ITRI, as we have seen, partners include IBM and Motorola (Freeman and Yue, 2001). The barrier for Scottish ITIs would arise if they do not systematically partner with TNCs such as IBM or Motorola. Second, such alliances are not principally centred upon basic research but development of already post-prototype technologies and as such are more about process efficiency and effectiveness than product development. The technology transfer and translation issues are thus less complex and swifter than those attending the kind of knowledge transformation likely to be conducted in ITIs. The barrier here for Scottish ITIs would arise if they do not attempt to directly commercialise near-market innovations, but instead concentrate on translating fundamental research into possible candidates for licensing (which then in turn may become a near-market innovation). Finally, such institutes work in consortia that include already existing technology-intensive SMEs or state-established technology leader companies (like CSM). In other words they do not engage in "academic entrepreneurship" commercialisation, which in Singapore, at least, occurs in Innovation Centres in privately owned Science Parks in the main. The potential barrier for the Scottish ITIs is that compared to countries like Taiwan, where such initiatives have been successful, there is not the same critical mass of receptive, technologically-advanced SMEs.

Nevertheless ITIs are a good example of positive thinking, creative adaptation and leadership in meeting an important element of the challenge set by the Scottish Executive in its "Smart, Successful Scotland" vision. This is

 GLOBAL KNOWLEDGE FLOWS AND ECONOMIC DEVELOPMENT – ISBN 92-64-10766-5 – © OECD 2004

the framework to which Scottish Enterprise works in the threefold strategy discussed above, linking the "Global Connections", "Growing Business" and "Learning and Skills" action lines.

Before exploring key elements of this strategy, such as optimising "FDI Spillovers", "SME Internationalisation" and "Improving Talent" in Scotland, it is worth dwelling momentarily on why ITIs encapsulate the heart of the management of the "International Knowledge Flows" conundrum.

Entrepreneurial and institutional regional innovation systems (ERIS and IRIS)*

A "Regional Innovation System", which is what Scottish Enterprise is embarked upon constructing, consists of two sub-systems. The first is the "Knowledge Generation" sub-system. The second is the "Knowledge Exploitation" sub-system. Most regions, and many nations, have poor linkage between the two sub-systems. Where nations or regions have overcome this barrier, it is either through successful working of market mechanisms, set in an appropriate regulatory environment, classically in the USA. Or market failure is overcome by the establishment of state entities that directly or indirectly straddle the "exploration" to "exploitation" divide. Scottish Enterprise is clearly embarked on the second of these to integrate necessary knowledge flows, since the first option is emergent but not yet mature. Indeed, by comparison with leading regional innovation systems in the USA such as that of Greater Boston and Massachusetts, or San Diego and Silicon Valley in southern and northern California, most European regions are constrained to public intervention if regional innovation is to function systemically. Of course, a majority probably do not have meaningful Regional Innovation Systems, something that contributes significantly to the much-touted "innovation gap" identified by the EU and others between Europe and the USA (Cooke *et al.*, 2000).

The key mechanisms facilitating the flow of knowledge, whether intra-regional, inter-regional or international, are knowledge itself, resources (particularly finance), and human capital. In strong market systems, venture capitalists that are proactive in seeking and assessing knowledge competencies in laboratories are crucial links across the exploration/exploitation boundary (Kenney, 2000). They are increasingly highly attuned to the nuances associated with specific, advanced fields of research, the "star"

* The ERIS (Entrepreneurial Regional Innovation System) and the IRIS (Institutional Regional Innovation System) represent a further elaboration of the regional innovation system concept referred to by Asheim in Chapter 2. The three fold categorisation used by Asheim could be further split into ERIS and IRIS to indicate that regional innovation systems could be more or less market driven or institutional based.

scientists associated with leading edge research, and risk assessment associated with its commercialisation. In systems such as Silicon Valley some scientists and engineers are highly attuned to stock markets, prospects for venture funding and initial public offerings (IPO). It is clear to see that the systemic nature of the likely interaction between scientific research, *i.e.* "knowledge generation" (itself involving exploration and *examination* of knowledge,the latter involving trialling and testing competences), and innovation or "knowledge exploitation" is massively assisted by these "boundary crossing" competences. To that must be added the prevalence of "academic entrepreneurs" managing a spinout firm while keeping an academic post in a nearby university, and receiving business management support from venture capital. These and their staff convey knowledge of distinctive kinds across boundaries too, and the micro-system of the firm operates as a seamless web. But added value comes from the fact that venture capital invests in portfolios of geographically proximate and non-proximate firms among which, at the inter-firm and inter-research centre levels, comparable knowledge transfer occurs both formally and informally. This occurs in biotechnology when an entrepreneur and a scientist sign an agreement to work on a definite project that is time-limited and in which monetary exchange occurs. Occasionally knowledge "spillovers" may be exploited where more informal discussion of less specific knowledge occurs. It is this network form embedded in market transactions and some "untraded interdependencies" that typifies the "open systems architecture" of the ERIS or Entrepreneurial Regional Innovation System that supports knowledge flows (Dosi, 1988; Best, 2001).

Where Entrepreneurial Regional Innovation Systems are underdeveloped, perforce ambitious regional administrations develop institutions to facilitate comparable effects through establishing "boundary crossing" institutions that may forge an "Institutional Regional Innovation System" (IRIS). The most fully researched of these is Baden-Württemberg in Germany, for which detail can be found in Herrigel (1996) and Cooke and Morgan (1998). To be brief, the key "boundary crossing" institutions are, for larger firms, the Fraunhofer Institutes, of which there are fourteen conducting applied research, and for smaller firms the Steinbeis Foundation, now numbering some three hundred transfer centres based in Higher Education Institutes and Innovation Centres. Fraunhofers conduct publicly subsidised, industry-funded research to solve technological or managerial problems, assess technologies and conduct foresight activities. They bridge the basic research function of the fourteen Max Planck Institutes and the ten universities across to the commercial application requirement from firms. Steinbeis provides a similar subsidised consultancy function for SMEs. Neither engages in spinout activity, which again is seen as an Innovation Centre or incubator function in Science Parks

GLOBAL KNOWLEDGE FLOWS AND ECONOMIC DEVELOPMENT – ISBN 92-64-10766-5 – © OECD 2004

located close to universities. Assessment of this version of an IRIS is that it works well where technology and innovation tends to be path dependent rather than disruptive (the latter being more typical of the ERIS set-up), where institutions have grown incrementally to meet needs in an evolving but well-understood sectoral innovation system (in this case automotive engineering), and where specialised in-house expertise familiar with technology application work is in place.

It is evident that ITIs (plus parallel measures already in existence, such as "Proof of Concept", the Alba Centre etc.) are intended to achieve outcomes comparable to those of an ERIS with machinery that is closer to that of a "stripped-down" IRIS (*i.e.* one without the full features of a structure such as that of ITRI in Taiwan). However, Scotland's IRIS escapes some of the "lock-in" problems that arise in the German case by having, in principle, competitive tendering for contracts that require "boundary crossing" research (applications) and licensing, spinout and sale of core IPR arising. Baden-Würrtemburg, for example, is an advanced engineering economy, and many institutes are "locked-in" to advanced engineering. It is very difficult to redirect these institutes away from their original mission in order to tackle new sectoral issues such as ICT or biotechnology research and commercialisation.

The intention to place many (though not all) of the applied research contracts in Scotland is commendable for competence building, but draws on a relatively small pool of expertise. Perhaps "lock-in" will still have to be guarded against by commissioning from outside Scotland. Correctly, and unlike the German IRIS exemplar, the fundamentals of "Knowledge In – Knowledge Out" mean that an unspecified, but presumably global as well as regional or UK "market" is driving ITI activity. This places a more considerable weight on "absorptive capacity" and "transceiver" capabilities on the part of some portion of a total of fifteen staff per ITI. In IRIS set-ups markets tend to be regional and highly attuned due to "institutional thickness" (meaning a multiplicity of innovation governance organisations). In ERIS contexts *network* organisational forms among firms reach global markets from a strong regional firm base of "many marketers". The GlobalScot programme will assist international knowledge flows in this regard, to some extent, although it is possible to interpret GlobalScot's current function as tilted more towards "Knowledge In" than "Knowledge Out". Nevertheless, it and the smaller, Alba Centre-focused Talent Scotland show recognition that "Knowledge" and "Networks" as well as "Markets" are inseparably intertwined.

To conclude this section, it has been shown that Scottish Enterprise's three-pronged strategy to optimise international knowledge flows to improve the knowledge economy capabilities of the Scottish economy and its enterprise support institutions rests on Scottish Executive's vision of a "Smart, Successful Scotland". This can be seen as a wise ambition given globalisation, entrepreneurial and knowledge economy trends visible clearly under current conditions of global free trade in liberalised markets. Scottish Enterprise's action lines targeting "Global Connections", "Growing Business" and "Learning and Skills" attempt, among other things, to create a more robust Regional Innovation System in Scotland. An innovative but key element of this is the ITI idea. This broad notion of an intermediary "boundary crossing" knowledge transformation institution moving scientific knowledge from its *exploration* to *examination* and *exploitation* phases has worked, albeit differently composed, in "Tiger" economies like Korea, Singapore and Taiwan and is, more elaborately, present in some of Europe's accomplished regional innovation systems, such as Baden-Württemberg. The Scottish approach is innovative but risky. It is innovative in seeking ERIS-type outcomes from an IRIS-type set-up without close alliance with leading technology corporations but with a vision of promoting possibly "disruptive" technologies and IPRs. The risk lies in this precise model never having been implemented hitherto; thus there is little guidance from experience as to its likely potential or pitfalls. It will be a success to the extent it induces support from all or a significant portion of institutions and firms in Scotland and beyond to which it is of consequence. As we shall see below, there are currently some grounds for believing that there is scope for substantial institutional network building before Scotland operates as a fully functional regional innovation system that might aspire to interact.

Overall, then, this section has argued that the best mechanism for stimulating knowledge flows is an Entrepreneurial Regional Innovation System (ERIS) typical of some of the most dynamic US regions, which is a network form embedded in market transactions, with some "untraded interdependencies". However, in the case of Scotland and most other OECD countries and regions the conditions required for a self-generating ERIS are not yet in place. Thus, we can argue that an Institutional Regional Innovation System (IRIS) is first needed in order to trigger development. Such an IRIS should not be too inward looking.

Types of knowledge in international knowledge flows

In this section, an attempt will be made to analyse, briefly and as clearly as possible, recent thinking about what might be termed "knowledge processing" rather than the more familiar "knowledge management" which has rightly been criticised by Hansen (2002) among others for its reinforcement of an outdated "command and control" organisational model. Knowledge processing implies consciousness of the importance of openness, inclusivity, transparency, tolerance and interaction as guiding precepts of the more heterarchical, less hierarchical model associated with successful knowledge intensive firms and organisations. Although much of what follows refers to set-ups that favour relatively unhindered knowledge flow of consequence among firms and relevant organisations, some of it is relevant also to intra-organisational flows, not so much of information and knowledge, but frames, references and schemes: in other words the *meaning* that transforms information into knowledge as the prelude to *action*. The language of frames, references, schemes and meaning derives from the Scandinavian "sense-making" school of organisational science (Weick, 1995; Ifvarsson and Häckner, 2002) that focuses on the use of management systems to bring to the firm's attention knowledge, its networks, and capabilities to exploit opportunities.

At the macro-level Qvortrup (2002) categorises knowledge as indicated in Table 4.1 in relation to the idea of optimising knowledge flows in regional innovation systems.

Table 4.1. **Knowledge categories and their meanings**

	Knowledge category	Knowledge category meaning
Concrete concept	Factual knowledge (*e.g.* qualifications, skills)	Knowledge of the world; codified, references
Analytical concept	Reflexive knowledge (*e.g.* competencies)	Knowledge of knowledge; partly tacit, schemes
Abstract concept	Systemic knowledge (*e.g.* paradigm knowledge)	Knowledge of the knowledge system; frames, assumptions, specific knowledge practices, may be tacit
Metalevel concept	Metasystemic knowledge (society as knowledge)	Knowledge of the conditions of the knowledge system; theories, understandings, meanings, actions, codified and tacit

To apply this framework to the issue under discussion, namely Scotland's policy and market engagement with the knowledge economy and its development of knowledge processing capabilities, we see how Scottish Enterprise's knowledge of the world, created by skilled interpreters of factual knowledge, rests on partly *tacit* schemes (or paradigm knowledge) and *assumptions* (tacit knowledge) that development now arises from *generative growth* (stressing socially interactive learning and networking, trust-building and collaboration for successful entrepreneurship) from within the knowledge system, that it is reliant on "bootstrapping" or focusing inward on current assets (*e.g.* GlobalScot; ITIs; Talent Scotland; Internationalisation of SMEs, Intellectual Property Rights etc.) as distinct from creating platforms for exogenous assets (Locate in Scotland/Scottish Development International; FDI; FDI supply chains; STAR Centres; etc.). Finally, at the metasystemic level, conventional wisdom is that the new paradigm cannot evolve satisfactorily in

Scotland without the significant intervention of Scottish Enterprise, and particularly the control centre at headquarters in Glasgow. This latter is reinforced by theories of market failure, peripherality, homogeneity (need for greater cosmopolitanism), small sized economy, and so on. Schematically, and for purposes of illustration only, it is not difficult to imagine different narratives especially at the metasystemic level that, on the one hand, might reframe the Scottish into a UK perspective and strategise around outstanding lobbying of the UK government (and EU) for resources, or reframe the market failure theory into one influenced by the idea of "crowding out" by the state causing weakening of market opportunity. Variations on these themes could then be worked back down the levels to the codified "Knowledge of the World", but it is fairly clear that nothing like this has been done in the interpretation of the "Smart, Successful Scotland" brief. Rather, a critic might argue that the latter has been "bolted-on" to a pre-existing schematic frame and set of meanings determined centrally.

One small, but telling instance of the effects of this occurred in respect of a question put by the review panel to policy makers in Scotland as to whether seeking to squeeze optimal value out of academic research by intensifying the quest for new businesses through academic entrepreneurship, assisted by the ITI institutions as well as pre-existing mechanisms supporting academic spinouts, "Proof of Concept" funding etc., was the best way to create innovative start-up businesses. This is not a question that aimed to be critical of that line of thinking; on the contrary the evidence suggests that within that "paradigm" Scotland performs above an (abysmal) EU norm. Rather, it was to assess whether other paradigms might not help achieve the goal of greater numbers of high quality, knowledge-intensive businesses in Scotland rather than substitute for the academic entrepreneurship plus ITI model assumptions. Discussion quickly turned to evidence-based policy thinking, the evidence in question pointing overwhelmingly to the superiority of *corporate* spin-outs being swifter growing, greater employers, and longer lasting entities than academic spinouts (see for example Lindholm Dahlstrand, 1997). Discussion ended with the rhetorical question from Scottish Enterprise: "Can you seriously expect a company to respond positively to a suggestion from a Scottish Enterprise executive that it lets five of its best people go to set up spinout firms? Oh! and by the way, we'd like five pieces of your IPR for them to take with them?"

The meaning of this is that within Scottish Enterprise headquarters there is a "Knowledge of the World" that is highly diplomatic towards larger, especially successful and foreign firms. A core assumption may be interpreted to be that one must be extremely careful not to do or say anything that *might* upset them or, as a worst case, question whether they might be happier somewhere else where provocative questions are less likely to be asked. This

GLOBAL KNOWLEDGE FLOWS AND ECONOMIC DEVELOPMENT – ISBN 92-64-10766-5 – © OECD 2004

is mostly tacit "conventional wisdom" rather than "Reflexive Knowledge". It is "Systemic (Paradigm) Knowledge" associated with the era of deference and "bend over backwards" to FDI. It is totally understandable that a responsible enterprise support agency should be exceedingly careful to do everything in its power to give good aftercare, be diplomatic and so on towards the companies it hosts. To repeat, observations such as this are not critique but an attempt at "sense-making".

Accordingly, the issue was raised with Scottish Enterprise executives in the LECs. This was an example of social scientific "triangulation" of observations to assess validity and reliability of apparent core assumptions. Three important insights were offered on the issue of potential variation in seeking to assist formation of new firms. First, generically, a view was expressed that despite presenting a notion of "communities of practice" for particular functions, relationships between Scottish Enterprise headquarters and the LECs in the field were not totally "joined up" but partly so, and that strategy from the centre and on the ground were not always linked up. This was perceived to be because "people are reluctant to give up pet projects or ways of working they are comfortable with". This, of course, can be said of officers at the centre or in the field, but it adds to the perception that the management of change to meet the vision of a "Smart, Successful Scotland" will benefit from close monitoring as it evolves.

Thus, a second important point made by field officers about the precise issue in question, was that their task as "account managers" and being in regular contact, including visits to firms in their area, gave them both a highly up to date view of firm trajectories, and open relationships with key managers. Because of this, account managers regularly discussed matters relating to knowledge about company strategy, including such matters as attitudes of firms towards corporate spinouts. Whereas Scottish Enterprise's Partnership Action for Continuing Employment (PACE) programme had been, sensibly, developed to advise on spinouts from downsizing firms (possibly signalling difficulties in specific markets) via, for example, a management buyout, account managers often discussed – without reluctance from corporate managers, wider opportunities for taking actions that might be interesting for the firm and Scottish Enterprise strategy. The third important point raised was that in such discussions firms were not unenthusiastic about the idea of a few good managers starting spinout companies. It happened occasionally, it was increasingly a feature of contemporary corporate strategy as competitiveness demands caused firms to out-source more rather than conduct the fullest range of requirements in-house, and if there were a more proactive position or a programme facilitating it, then more corporate spinout activity might be forthcoming. In other words, through being regular connectors with firms on the ground, the frame of reference of field officers was different from, more contemporary than, and possibly more in tune with Scottish Enterprise goals than the predominant frame of reference at headquarters. So a better knowledge management process would bring grassroots knowledge into the strategic centre so that policy making would be better informed and more up to date.

How might such disconnects be moderated? The question arose of an enhanced Intranet facility within Scottish Enterprise, parts of which would be devoted to LEC-Headquarter (HQ) interactions of the kind discussed. It would need time to populate it with requisite information, and commitment from field officers and HQ staff to keep it updated. Moreover job descriptions would need to specify as a requirement that it should be an everyday responsibility to find out what of relevance was posted upon it. To give weight to the importance of intra-organisational "Knowledge of the World" this knowledge management and processing function should be given status through the appointment, presumably at the centre but with network-wide scope, of a senior Knowledge Manager. The model provided by GlobalScot as an Extranet, albeit focused on Scots and Scotophiles abroad, is a good one upon which to build such functionality. Ironically, one of the greatest potential value-adding functions of GlobalScot, which would be to tap into the knowledge being communicated by persons with valuable anticipatory or foresight understanding of market signals was not designed in from the outset because, correctly, of confidentiality concerns. But a polling function to access key international knowledge (including perceptual as well as explicit knowledge) flows while maintaining security is needed and reported to be in development at Scottish Enterprise. Hence intra-organisational learning clearly creates opportunities for developing wider Reflexive Knowledge competencies within the organisation and between it and its customers. In the process that can assist Systemic or Paradigm Knowledge development, making transparent tacit assumptions that may be unreliable.

Knowledge flows in life sciences as an exemplar

Mention of the kind of *anticipatory* international knowledge flow likely to be circulating on the GlobalScot network draws attention to the importance of many kinds of knowledge in an economy that has become increasingly knowledge-intensive, notably of the scientific kind. In fact, a most perceptive insight into a possibly crucial aspect of another area of "disconnect" in Scotland's regional innovation system is provided by Bruno Latour:

> "Science is certainty; research is uncertainty. Science is supposed to be cold, straight and detached; research is warm, involving and risky. Science puts an end to the vagaries of human disputes; research creates controversies." (Latour, 1998, cited in Nowotny *et al.*, 2001)

 GLOBAL KNOWLEDGE FLOWS AND ECONOMIC DEVELOPMENT – ISBN 92-64-10766-5 – © OECD 2004

He identifies key distinctions between "science" and "research". Science is conditioned by relative stability, both in its historical evolution and its Systemic Knowledge, the latter not least because it has to be taught and learned according to certain frames of reference that, for practical reasons, cannot constantly be required to be in flux. Thus to some extent, science is always somewhat out of date. Research, by contrast, is the force undermining scientific stability. It is dynamic, questing and questioning; it often works at the margins of uncertainty and sometimes by backward extension brings about fundamental change to the core of a paradigm. Scottish universities, like others, may not have examined their "Knowledge of Knowledge" reflexively enough in such a way. Thus they can evidently be perceived as "stand-offish". Occasionally, as Kuhn (1962) noticed, broader paradigm shift can occur. One of the best examples of that is the so-called "molecular biology revolution" that heralded the arrival of biotechnology in the 1970s but only became fully worked through in the 1990s. Briefly, three major things attend this revolution, each of which spread far beyond "science" itself into economic affairs, signifying the power of "research" itself as the motive force of the knowledge economy and society (Nowotny *et al.*, 2001).

The three are the following. First, since its origins as a modern industry, which may be dated to the Bayer company's synthesis of aspirin in 1893 (though its properties were known by Hippocrates and Galen), pharmaceuticals has been a province of synthetic chemistry with pharmacologists seeking small molecule inhibitors of disease. Molecular biology brought the beginnings of understanding of large molecule composition at the level of genes and proteins that are not susceptible to synthetic chemistry. Second, even the largest and most successful pharmaceuticals firms remain essentially synthetic chemistry firms, professionally "locked-in" to the frames and methods of that discipline and weak in their internal R&D capabilities in molecular biology. Thus they are increasingly dependent on partnerships with dedicated biotechnology firms (DBFs), many of which are institutional or corporate spinouts that also tend to cluster geographically. Third, the key source of "life sciences" knowledge of great potential relevance to modern pharmaceuticals businesses and ultimately patient healthcare through national health services is laboratory research in specialised centres or departments in universities and other research organisations.

Indirectly or directly, this knowledge reaches patients as drugs via pharmaceuticals firms, and nowadays some DBFs of requisite scale. That is "pharma" may link direct to research laboratories or be intermediated by single or, more likely, networks of DBFs. Scottish Enterprise and the Scottish Executive, not least through its *Science Strategy for Scotland*, are acutely aware of this, have for many years put in place enterprise support mechanisms to assist "boundary crossing" for research into commercial innovation, and have displayed considerable success. So much so that Scotland is the UK's third key biotechnology location after Cambridge and Oxford. But Scotland seeks to do better, finds the successes slow in reaching fruition, and the critical mass believed to be crucial to "take-off" absent. Correctly, it is seen that enlarging the talent pool by holding on to Scottish-trained talent and augmenting it from outside Scotland as well as devoting more training and research resources to creating it in the first place is necessary. It is thought that part of the "critical mass" problem includes a relative absence of large pharmaceuticals ("big pharma") firms in Scotland. Expecting production in the few that are present quickly to move offshore to cheap labour zones, part of "Global Connections" involves attracting corporate headquarters and R&D facilities, not only in life sciences but in other targeted sectors.

In consultation with one of Scottish Enterprise's success stories in fulfilling that ambition – Organon, the US-based part of Dutch pharmaco-chemicals company Akzo-Nobel, located at Newhouse, it became clear that past thinking about building integrated supply chains to serve such firms may be redundant. Organon has, with Scottish Enterprise and Executive assistance, closed production but opened an eventual 350 personnel basic research facility specialising in Psychiatry with Arteriosclerosis, and Anaesthesiology. It is in Newhouse, central Scotland, partly because of path dependence deriving from its arrival in 1948 and its loyal workforce (which nevertheless must change with the move to basic research), but particularly nowadays because of its location in reasonable proximity to top life sciences research laboratories and talent in Glasgow, Edinburgh, Dundee and St. Andrews universities.

Three important points were made about contemporary life sciences research dynamics of relevance to a "Smart, Successful Scotland" ambition. First, Scotland produces 20% of UK biotechnology graduates, 28% of those in medicine and 18% of bioscientists. But Organon is the only pharmaceuticals research company. Improved "Knowledge Out" regarding the existing potential talent pool would help alert other such businesses to Scotland's attractions. Second, to achieve targets for increasing R&D spend and employment in Scotland, it should be recognised that Organon spends 17.5% of sales revenue on R&D – exactly the industry average worldwide. That is twice the average in any other sector. A view that Scottish Enterprise might consider is that targeting pharmaceuticals R&D with a focused strategy would bring better returns for the same investment in raising Scotland's Business Expenditure on R&D (BERD) statistic, which is currently sufficiently low to be a serious cause for concern. Third, while critical mass is important, supply chain promotion might not bring major benefits in life sciences in the short term, since, at the leading edge there are, globally, critical mass shortages in many areas. In the longer term, however, if critical mass can be mobilised it will create a virtuous circle of the kind to be described in the last paragraph of

GLOBAL KNOWLEDGE FLOWS AND ECONOMIC DEVELOPMENT – ISBN 92-64-10766-5 – © OECD 2004

this section. Thus Organon needs to buy in expertise in target compounds, genomics and assay development which, for its purposes means dealing with Californian firm Incyte, which has the only genetic database for bioassays. All firms everywhere with comparable needs must deal with Incyte, and were there a local firm it might well not get the business because Incyte is the best.

Most Organon research collaborations are in the US where the biggest institutes of relevance are found. This too is a function of sector maturity and the ten-year lead of the US over the UK in biotechnology.

Now, finally, what emerges about apprehending international knowledge flows in life sciences that are of wider relevance to the "Smart, Successful Scotland" vision both for life sciences and the broader sectoral picture regarding the Knowledge Economy and engaging as fully as possible with it? Three key points can be made. The first is that markets are remarkable mechanisms incentivising firms to identify, as Organon's case shows, the single source of vital knowledge to it in the world. There is no mystery in this; without such knowledge, the firm would be less effective and might risk ceasing to function. Scottish Enterprise can learn from its contingent of life sciences firms how it could better mimic their apprehension of vital knowledge and integrate such a service either purchased from knowledge market providers or developed in-house to raise the knowledge network capacity of the industry. Second, though, it is likely that research-intensive firms in biotechnology, ICT and new media gain useful knowledge from "knowledge spillovers". These can be pecuniary, just as Organon pays for knowledge spillovers from good Scottish life sciences training by accessing human capital. Or, maybe to a lesser extent (although some literature on "untraded dependencies" *e.g.* Storper, 1995, argues more) is non-pecuniary, tacit, "heard in the restaurant" in kind. Clusters are said to offer significant added value from the "surprise" knowledge that emerges from the interaction of distinct tacit and part-codified knowledge exchanges they foster. Scotland has a notable cluster policy, including biotechnology. However, it was possible to detect a slight cooling in enthusiasm for clusters, from conversations with Scottish Enterprise officers, perhaps because of slow, limited effects and, understandably, due to many other regions and nations having followed suit. Clusters offer one important means of building critical mass and the amplification of knowledge spillovers of both kinds. Because, thirdly, as experience in the main clusters in the US shows, particularly the leading, more basic research-led one in Greater Boston, "ahead of the curve" knowledge coming from the likes of the Whitehead Institute, where much of the human genome was decoded, brings in, in short measure (between 1999 and 2002), "big pharma" like Novartis, Pfizer, Abbott Labs, Wyeth and biotechnology leaders like Amgen to manage their basic research in a good research climate. Although Novartis announced a $25 million deal with

University of California, Berkeley to have first access at its patents, in 2002 a £250 million Novartis Genomics Institute investment went to Cambridge, Massachusetts for obvious reasons.

Hence, in summary, life sciences could be a jewel in the crown of a "Smart, Successful Scotland" if existing policy is broadened from the narrower biotechnology support policy, present strengths in human capital are better-marketed but also augmented, and the cluster strategy re-invigorated for a broader life sciences and, particularly, pharmaceuticals research constituency. The whole could usefully be underpinned by a state of the art international knowledge flows network management system based on open access to a service designed in partnership with industry and academia.

The Scottish universities also need to be pulled into an even stronger position as sources of disruptive change through research rather than cosy purveyors of established science. The policy implication of this is that a consolidated effort to build up the Scottish research base, including institutes, talent and funding is needed. This is not simply a question of resources for near market innovation but also "ahead of the curve" research.

An examination of policies in other countries also offers some ideas for the development of policies to promote life sciences in Scotland. Briefly, other, stronger knowledge economies that specialise in life sciences do three things that do not appear to occur in Scotland. In Sweden, the Regional Innovation Systems Agency (VINNOVA) regularly monitors the bibliometric measures of Swedish bioscientists, logging and publicising information on "ahead of the curve" research and publication, and with which institutes Swedish scientists are collaborating on this. Second, in North Carolina where some 142 biotechnology firms have evolved, a Regional Biotechnology Institute houses many small research arms of global "big pharma" (Aventis, BASF, Bayer, Eli Lilly, and GlaxoSmithKline) in the same building as new spinouts so that knowledge flow and interaction can be optimised. Also in North Carolina, the State University's Centennial Campus houses research laboratories on campus next door to the equivalent university laboratory to enhance "boundary crossing" knowledge flows. Finally, in the USA, the EPSCoR programme (Experimental Program to Stimulate Competitive Research) gives research grants to lagging regions by requiring a lower approval standard for bids for research funding. This helps build up research excellence in fields that are not being focused on by the major centres and brings opportunities for inward and outward knowledge flows to more places.

Will the Scottish Enterprise Strategy work and what fine-tuning might assist it?

There are strong reasons to believe that the broad strategy on which Scottish Enterprise has embarked is highly appropriate if it is correct to assume (Systemic Knowledge/Paradigm Change) that we are already in a "Knowledge Economy". The indicators discussed in the introduction to this chapter and others advanced by authors like Dunning (2000) offer convincing evidence that we are. Moreover, there is little to be gained by pursuing the kind of FDI production platform strategy that served Scotland well in the past, under conditions of globalisation. Scottish Enterprise is pioneering a paradigm shift within its own organisation by seeking to build an enterprise support system attuned to the uncertainties posed by research-driven economic activity of the kind intimated at by Latour. It is correct to build up capabilities for *Generative Growth* from indigenous resources, and to promote a networking and learning culture intra-organisationally for Scottish Enterprise and externally for international knowledge flows among firms and institutions that generate and exploit knowledge of consequence to Scottish economic prospects.

However the headline question asks will the strategy of "Global Connections", "Growing Business" and "Learning and Skills" work? That is, will it deliver a "Smart, Successful Scotland'? There are three reasons why it might not. First, the strength of pursuing what is called here a *Generative Growth* strategy, but which may without careful thought be unreflectively equated with "endogenous growth" may, as we have indicated, pay more attention to individualistic "entrepreneurialism" rather than a more networked, systemic innovation model would warrant. This returns us to the possible conflict between ERIS and IRIS. Much of Scottish Enterprise's ambition is motivated by observation of the rapid mobilisation of new business growth in strongly marketised, largely private enterprise models like Silicon Valley. We have seen that Scotland is not cognate with such a system. Markets tend to fail parts of the UK like Scotland in a national system in which London and its region dominate the UK's knowledge economy activities to such an inordinate extent. There is thus more of a case for a kind of "collective entrepreneurship" animated by intermediary institutions where existing institutions are seen to be underperforming or market forces absent. Clear thinking about and learning from probable setbacks in the initiation period of the ITIs, for example, will be important. This is not least because the first will be entering the unknown, attempting not to transfer established knowledge into receptive SMEs but translating, at arm's length, basic research into commercial innovations with spinout firms a possible vehicle for commercialisation. Knowledge of "the market" will need to be highly refined if returns are to be successfully accrued. This is because, not least, "markets" are capricious. Strategy needs to be based on a longer-term view than "markets" normally offer. This does not mean a return to "picking winners" but rather weeding out obvious losers from a consensus based strategy of what is both needed socially and can be exploited commercially.

Second, there is already an interesting, innovative but ultimately limited trend towards focusing inwardly to exploit "endogenous" assets in many of the action lines being conducted. One might be forgiven for interpreting strategy as recoiling somewhat from the disappointments of unfulfilled FDI expectations, and investigating, almost as a *terra incognita,* Scotland's own, possibly under-utilised, economic assets. Thus, seeking (in the case of Organon successfully) to substitute R&D for production *in situ* may be more difficult generically, where as we have seen R&D expenditure is universally lower than in pharmaceuticals. GlobalScot, and to a lesser extent, Talent Scotland are good prototype electronic networking instruments that exploit Scotland's own resources. But these are inevitably limited, and currently synergies arising from them remain secret, in the case of GlobalScot, from the organisation that facilitated it and could add value to its enterprise support services given a means to process and disseminate the most useful elements of it. Having said that, such modest pioneering actions give Scottish Enterprise experience and a lead in follow-up, more systemic knowledge management and processing methodologies and envisioning of appropriate tools for future development.

Finally, two related sub-points. The Knowledge Economy is highly globalised. Research-intensive firms must interact with specialists of many kinds in many different and distant countries. Yet, the globalised world economy operates through global value chains or global value networks (Dicken *et al.*, 2001) that focus at ground level on regional and local clusters. We saw how pronounced a feature of the life sciences and biopharmaceuticals industries such an organisational framework has become. But it is true also of ICT, software (think of Israeli specialisation in global sales of Internet "firewalls" from Herzliah between Tel Aviv and Haifa), and the global media cities like London, Los Angeles, New York, Munich and so on. So the prospects for building comprehensive chains of supply regionally are less. Nevertheless, the importance of having clusters of competing and collaborating firms in leading edge sub-sectors of such industries becomes more and more important. But strategy needs to be scaled up from, say, biotechnology clusters to life sciences or healthcare innovation systems and required resources for basic research appropriated. The obverse and second sub-point is that the global audience needs constantly to be connected to the qualities present in Scotland, notably to return to life sciences, the hugely above average production of life sciences talent in Scotland that is itself a potential attractor of pharmaceuticals R&D. Moreover, heightened focus on the 17.5% of sales

 GLOBAL KNOWLEDGE FLOWS AND ECONOMIC DEVELOPMENT – ISBN 92-64-10766-5 – © OECD 2004

habitually spent by pharmaceuticals firms on research would represent a better rate of return from one new "pharma" than two new ICTs. So, in conclusion, Scottish Enterprise will have to develop its own internal Reflexive Knowledge (self-understanding) to further adjust its Knowledge of the World (problems and policy solutions) to match the new Systemic or Paradigm Knowledge (new knowledge economy) with which it has begun to engage.

Now, briefly, some fine-tuning possibilities have been alluded to. These include efforts to improve internal knowledge management methods and tools within Scottish Enterprise the better to hear what Knowledge of the World the field offices bring but also to better synchronise communities of practice from field to centre. A modern database-driven, knowledge management expert system is imperative here. Second, extension of that capability to the external world of "knowledge of consequence to knowledge-intensive firms and institutions" would be a major advantage for Scotland. This implies extension of the "expert system" outside the limits of Scottish Enterprise. Third, keeping but broadening and deepening the cluster and network concept to encompass the concept of Scotland as a Regional Innovation System would clinch the joined-up aspiration institutionally in important ways by improving internal and external knowledge flows. Fourth, recognising the future likely importance of healthcare and targeting R&D attraction and development in relation to known "talent" assets will help meet BERD targets more efficaciously. Finally engaging internally in a Reflexive Knowledge analysis will be useful as an organisational learning process that will assist in the jettisoning of "Industrial Age" pet projects that accrete to all enterprise support systems.

Conclusions

This contribution began by itemising the three key pressures to which Scottish Enterprise's new Knowledge Economy approach is a response. These were *globalisation* which, it is now evident, favours movement of the kind of economic activity that assisted Scottish economic restructuring in the past to Central and Eastern Europe, Asia, Africa and Latin America. Secondly, *entrepreneurship,* that rises up the policy agenda as FDI becomes harder to attract. But entrepreneurship is not uniform as a skill within society, nor something that individuals alone can fulfil ubiquitously. An entrepreneurial social environment needs to be created consisting of, ideally, a great variety and choice of valued services in the market, but where markets do not function adequately great care has to be devoted to provision of access to appropriate alternatives. Earlier mention of Israeli "firewalls" expertise draws attention to the development blockage caused by absence of venture capital for potential spinouts, the creation in 1991 and privatisation in 1997 of a public venture capital vehicle, and its success in both linking start-ups to US venture capital, and attracting them in turn to open offices in Israel. Some 3 000 start-ups flourished in security software in Israel during the 1990s as a consequence of this. Finally, the *knowledge economy* itself is a new kind of constraint on traditional enterprise support capabilities, as we have seen.

Above all it seems to demand vast increases in public budgets for university research and for training the "talent" on which it ultimately resides. Keeping and attracting talent while augmenting budgets demand focus and prioritisation of a kind that development agencies traditionally can feel uncomfortable about. But it is impossible for even much larger economies than Scotland's to be globally competitive in all sectors, so the challenge is to transcend the old "picking winners" approach by a "commitment to vision" approach that Scottish Enterprise has shown itself relatively well-attuned to in the recent past.

Then analysis was provided of the new mechanisms being introduced to assist Scotland's economy to fulfil the ambitions set for Scottish Enterprise through the Scottish Executive's aspiration for a "Smart, Successful Scotland". The ITI idea received considerable attention both in terms of its possible role models and the differences between the idea of an intermediary technology body in the Scottish context and those that had been important to the rise of the Asian "Tiger" economies. If it works, the ITI approach could remove an important Knowledge Economy blockage in Scotland, which is the rising but still inadequate number of new knowledge-intensive businesses that may assist in keeping Scotland's working age population within a modern Scottish labour market. Something of a contradiction between the entrepreneurial aspiration and the institutional methodology was identified. The "stretch" from translating basic research across the boundary to commercialisation is a large one in the absence of the kind of "institutional thickness" present in certain regions in more "corporatist" economies like Germany. But that some institutional innovation is needed to achieve *generative growth* is undeniable and it will be interesting to see how ITIs, when established, learn the necessary crafts.

An approach to thinking about the relationship of an enterprise support organisation and its constituency in the Knowledge Economy was outlined in which different kinds of knowledge are seen to benefit from being consciously reflected upon to reveal hidden assumptions, conventional wisdoms and frames of meaning that may hinder the achievement of progress in conditions of radical paradigm change such as that attending the rise of the Knowledge Economy. It is likely that some of the disconnects inside Scottish Enterprise and between it and some newer constituents like universities may come about because of a certain absence of Reflexive Knowledge and Paradigm Knowledge at various points in what, were these disconnects to be moderated, could be a better functioning, more "joined-up" regional innovation system. Finally the

 GLOBAL KNOWLEDGE FLOWS AND ECONOMIC DEVELOPMENT – ISBN 92-64-10766-5 – © OECD 2004

question as to whether the strategy adopted would work was posed and fine-tuning issues raised. A number of moderate but, it is hoped, relevant comments were made that are intended to help move Scottish Enterprise ever closer to its aim to improve its appreciation, management and dissemination of knowledge internationally and within Scotland to enhance the capabilities of Scottish firms and institutions to compete globally. These include: improving internal knowledge flows management by establishing an appropriate "expert system"; improving the quality of knowledge available inside Scottish Enterprise and outside to its constituents and customers by undertaking better knowledge-focused activities, such as bibliometric audits, co-location of corporate R&D and spinouts, and research affirmative action (*cf.* EPSCoR); developing Scotland's regional innovation system capabilities by a thoroughgoing linkage of knowledge exploration and exploitation competences; placing a greater emphasis and understanding of healthcare and life sciences as a carrier of future global innovation opportunity; and conducting an institutional self-analysis process to fully embed the requirements of a knowledge economy in the institutional set-up at Scottish Enterprise.

References

Arthur, B. (1994), *Increasing Returns and Path Dependence in the Economy*, Ann Arbor: Michigan University Press.

Best, M. (2001), *The New Competitive Advantage*, Oxford: Oxford University Press.

Cooke, P. (2002), *Knowledge Economies*, London: Routledge.

Cooke, P. and Morgan, K (1998), *The Associational Economy*, Oxford: Oxford University Press.

Cooke, P., Boekholt, P and Tödtling, F. (2000) *The Governance of Innovation in Europe*, London: Pinter.

Department of Trade and Industry (1999) Biotechnology Clusters, London, DTI.

Dicken, P. *et al.* (2001), Chains and networks, territories and scales: towards a relational framework for analysing the global economy, *Global Networks*, 1, 89-112.

Dosi, G. (1988), Sources, procedures and microeconomic effects of innovation, *Journal of Economic Literature*, 26, 1120-1171.

Dunning, J. (ed.) (2000), *Regions, Globalisation and the Knowledge-Based Economy*, Oxford: Oxford University Press.

Flora, J., Sharp, J and Flora, C. (1997) Entrepreneurial social infrastructure and locally initiated economic development, *Sociological Quarterly*, 38, 623-645.

Freeman, C. and Yue, C. (2001), The case of Singapore, in C. Yue, N. Freeman, R. Venkatesan ajd S. Malvea, *Growth and Development of the IT Industry in Bangalore and Singapore*, New Delhi: Sterling.

Hansen, M. (2002), Turning the lone star into a real team player, *Financial Times*, August 8.

Herrigel, G. (1996), *Industrial Constructions: the Sources of German Industrial Power*, Cambridge: Cambridge University Press.

Ifvarsson, C. and Häcker, E. (2002) Strategic knowledge management; a sense-making framework for control of spillover effects in interorganisational networks, University of Technology, Luleå (mimeo).

Kenney, M. (ed.) (2000), *Understanding Silicon Valley*, Stanford: Stanford University Press.

Latour, B. (1998), From the world of science to the world of research? *Science*, 280, 208-209.

Lee, C. (ed.) (2000), *The Silicon Valley Edge*, Stanford: Stanford University Press.

Lindholm Dahlstrand, A. (1997), Entrepreneurial spin-off enterprises in Gothenburg, Sweden, *European Planning Studies*, 5, 659-673.

Mathews, J. and Cho, S. (2000), *Tiger Technologies*, Cambridge: Cambridge University Press.

Morrison, C. (2002), Sectors – electronics, Presentation to Scotland in a Global Economy: the 2020 Vision, Glasgow, November 25.

Nowotny, H. *et al.* (2001), *Re-Thinking Science: Knowledge and the Public in an Age of Uncertainty*, Cambridge: Polity.

Organisation for Economic Co-operation and Development (1994), *The OECD Jobs Study*, Paris: OECD.

Porter, M., Sacks, J. and Warner, J. (2000), *The Global Competitiveness Index*, Boston: Harvard Business School Press.

Qvortrup, L. (2002), The region as a developmental knowledge laboratory, Presentation to International Workshop on Digital North Jutland, Aalborg, November 21.

Romer, P. (1990), Endogenous technological change, *Journal of Political Economy*, 98, S70-S72

Storper, M. (1995), The resurgence of regional economies, ten years after: the region as a nexus of untraded interdependencies, 2, *European Urban and Regional Studies*, 191-221.

United Nations Industrial Development Organisation (2002), *Competing Through Innovation and Learning*, Vienna: UNIDO.

Weick, K. (1995), *Sense-making in Organisations*, London: Sage.

Weissmann, G. (2002), *The Year of the Genome*, New York: Times Books.

ISBN 92-64-10766-5
Global Knowledge Flows and Economic Development

Chapter 5

The innovation solution: some lessons from Scotland, theory and Canadian cities

by

John de la Mothe (University of Ottawa, Canada)

This chapter argues that cities and regions should follow strategies of "constructed advantage" to engage with the global knowledge economy. Constructed advantage strategies involve i) attraction and retention of talent – requiring business development opportunities, risk capital availability and market access as well as quality of life, ii) attraction of investment – requiring access to smart people, infrastructure, funds, technologies, markets and a complementary regulatory regime, and iii) appropriate public policy – requiring leadership and collaborations from all stakeholders. Examples are given from Canadian cities.

"Interactive learning and collective entrepreneurship are fundamental to the process of innovation." B.Ä. Lundvall (1992)

Introduction: some observations of the Tartan Tiger[1]

Scotland is deeply enmeshed, as are many other advanced small and medium sized economies, in an important transformation as they enter the global knowledge economy. Just since 1997 (1997 = 100), Scotland's Gross Domestic Product has grown to 113.3. However, agriculture, forestry and fishing have rested at 99.3. Production is down to 96. Construction has grown, throughout various business and investment cycles, to 106.1 while services – largely in such knowledge-intensive activities as ICT – have impressively outpaced other sectors by an indexed growth of 122.1 (Q4, 2002).[2] GDP per head is currently US$22 000 as compared with Canada which stands at US$26 000.

With a population of 5 million people, 30% are engaged in managerial or professional occupations. Unemployment has dropped to 4% (or just over 6% using the International Labour Office measure) yet 13% of the unemployed have been out of work for a year or more. The distribution of population is highly variable, with as many as 3 300 people per square km. in Glasgow and only 8 in the Highland Council Area. And Scotland is declining, with the highest migration rates among the highly skilled/qualified. Scotland has an international reputation for excellence in higher education[3] and since 1998 enrolments in ICT fields has increased by 45%. Scotland is home to 20% of UK biotech start-ups, and is the 6th largest equity market in Europe (managing about £350 billion in funds). But between biotech, optoelectronics, telecommunications and semiconductor fabrication, Scotland employs only 37 000 people (out of a labour force of 2.5 million).[4] Sixty-five per cent of Scotland's exports flow into the European Union, but the top service exports arise from non-knowledge intensive activities, such as tourism (27%) and oil and gas (23%), followed by banks, insurance, assurance, and higher education (32%). Computer and software services account for 18%. Thirty eight per cent of manufactured exports come from Scottish office machinery.[5] However non-Scottish firms – such as Cisco, Motorola Lucent, IBM, Hewlett Packard and Compaq – dominate the Scottish tech sectors and manufacturing. They tend to be at the lower scale of production and towards the lesser value-added end of the R&D spectrum. Official Scottish documentation notes positively an

 GLOBAL KNOWLEDGE FLOWS AND ECONOMIC DEVELOPMENT – ISBN 92-64-10766-5 – © OECD 2004

ability to attract over 200 call centres, even though these have been shown to be highly mobile, low value-added, low paying and lacking in regional staying power. Scottish productivity trails Finland, the Netherlands, Belgium, Germany, France and the United States. Its entrepreneurship index shows lower levels than the UK, and digital connections are two thirds those of London.[6]

Thus it can be succinctly stated that Scotland faces a number of issues as it transforms into a knowledge economy. The central challenges are related to how Scotland can:

a) attract talented labour,

b) retain talented labour,

c) optimise knowledge spill-overs and value creation, and

d) attract high value-added foreign direct investment.

In order to achieve this, Scottish Executive and Scottish Enterprise have designed a number of policies and programmes geared towards addressing these issues. It has a series of complementary activities, such as *Scotland: A Global Connections Strategy* (2001), *A Smart, Successful Scotland* (2001) and *A Science Strategy for Scotland* (2002). These documents illustrate the good progress that has been made since devolution.

However, major challenges remain. The science strategy bears a remarkable and undifferentiated resemblance to the science strategies of a number of countries, including Canada, South Africa, Finland, Denmark, the Netherlands, and Australia. It lacks specificity to Scotland's unique capabilities, opportunities and challenges. It emphasises schools and public awareness, but does not locate itself within the economic context of the entrepreneurial, innovative, knowledge-intensive future of Scotland. Direct comparisons are problematic in that independent data for Scotland (*à la* OECD's Frascati and Oslo Manuels) are not yet available, looking for example at GERD/GDP, HQP/capita, and so on[7]. Moreover, there seem to be major disconnects in the production and flow of value-added knowledge. For example Scottish universities, which are world renown, also show disinclination towards collaboration with local players, especially other Scottish universities. International bi-lateral relations seem to be greatly preferred over multi-lateral research with local industry. Note Figure 5.1.

Scottish innovators rely more on distributed supply chains – notably distant final customers and on local suppliers for innovation, but not on local competitors, collaborators or universities.

As a result, the potential for building scale effects which are often required for competitive, world playing (never mind world leading), research is largely missing. A similar tendency can be seen in the private sector, where

Figure 5.1. **Spatial distribution of innovation-related linkages**

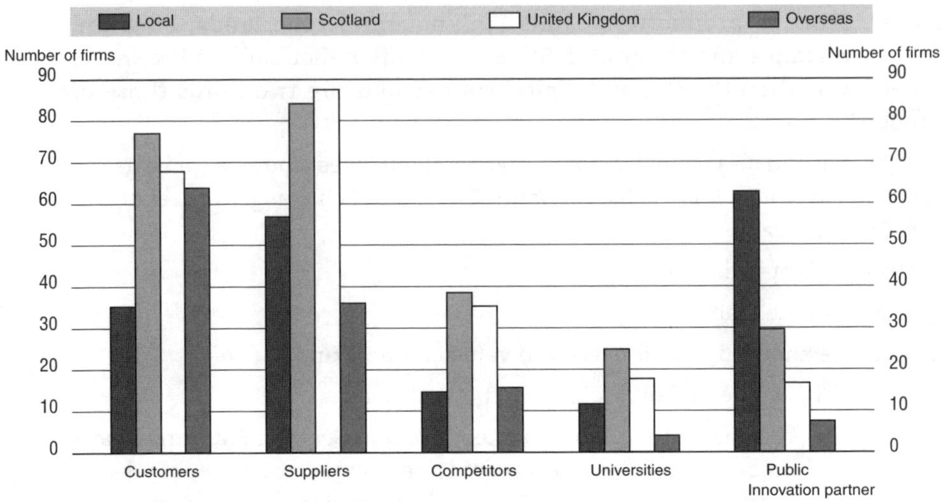

Source: Scottish Enterprise.

pharmaceutical firms (for example) situate themselves in stand-alone sites (not in science parks or knowledge communities) and deliberately separate their research, development and design functions, only to distribute the value-added components abroad – still formally within the company – but with the sales, development and profit activities accruing to other European regions – not Scotland.

These types of examples, all gleaned from interviews with university officials, private sector executives and public servants in Scotland by the OECD Expert Panel, point to an important series of observations. One is that in the 21st century knowledge economy, Scotland cannot rely on low value-added manufacturing and services, or traditional cultural industries, in order to sustain a high value-added economy. Nor is isolation an option. It must develop high value-added activities and engage the global economy. In order to do so it must embrace the economic dynamics of knowledge and innovation. It must adjust its longstanding commitment to a neo-Ricardian view of trade and production that is based on comparative advantage in which "value-added" is defined in terms of "what we have". And given Scotland's small domestic market base and population size, it cannot follow a neo-Porterian competitive advantage strategy, which centrally assumes scale and scope of either domestic consumer market or industrial capacity. Instead, Scotland might well consider following a strategy of constructed advantage. What might this mean?

Some lessons from theory

In the global economy, much popular literature has emphasised the growing importance of interdependencies between regions and the speed of interaction because of information and communication technologies. Indeed, many of the citizens of OECD countries – including Scotland – are now "on-line" both at work and at home.[8] They're connected. Scotland and Canada are both nations that claim policy and government programme success in connecting their citizens to the Internet. But connectivity in and of itself does not lead to regional economic development.

In addition, influential contributions to this literature have also suggested that the nation state is dead (Guehenno, 1995 and Ohmae, 2000). This is a non-sense. Governance is indeed more complex, involving the play between local institutions (firms, colleges, universities, city halls, local councils, etc), regional governments, national governance bodies (including Non-Governmental Organisations) and multinational agencies such as the European Union (EU) and the World Trade Organisation (WTO). But governance now is also *more* important (de la Mothe, 2002). Traditional arguments about industrial policy and top-down government intervention in the market place are now moot. Scotland knows this, given its on-going devolution. But new arguments about innovation policy through which local ingenuity, entrepreneurial vigour and appetites, reach up and are met by regional and national government policies and programmes which are adaptive enough to in essence become customised to local needs is now the way forward. Why is this?

Growth

Economic growth is local. National aggregate statistics notwithstanding, the causes – and benefits – of sustainable economic development are embedded in local institutions and people. In other words, if the OECD estimates that a country will experience growth of say, 4% next year, no one expects this growth to be evenly distributed across every region. Growth is therefore "lumpy". Foreign investments, industrial concentration and talent agglomerate in areas that have prepared and culturally conducive institutions. We have seen this repeatedly in empirical studies, from comparisons of Route 128, Silicon Valley and science parks to numerous assessments of clusters and cities such as Dresden, Singapore and Austin (Sassen, 2000).

Location

In all of these studies, the readiness of local and regional economies has proven to be decisive. The role of robust and active governance – a matching of "top-down" policy making and programme design with "bottom up" leadership and action – is key. Governance is no longer about picking winners but is one of backing leaders! Of course, on one level, this simple observation is not surprising given Marshall's work, a century ago, on industrial districts. In this important and well known work, the decision to locate a firm, to start a firm, the demand and supply of skilled labour, the draw on local and foreign investment capital, the inculcation of entrepreneurial drive, and so on, all relating to traditional factors of production... were seen as key. This is still true. Manufacturing and primary services still matter. Indeed, these combinations of factors of production allowed trade patterns, based on local advantages such as natural resources, to be well understood. "Value-added" was based on what was "at hand" (an abundance of wood, marine life, coal, climate, and so on). Much of this, which can be noted here for its emphasis in terms of improvements of manufacture, improvements of technical skills, the development of local value chains, market access and trade profiles, can be linked with the work of the great Scottish political economist, Adam Smith.

Factors of production

But what has dramatically changed since the time of Smith and his "invisible hand", has been a global and critical shift in the factors of production. No longer do we rely, either analytically or in terms of strategy, decision and policy, on simple capital and labour ($Q = f(KL)$) equilibrium. Instead, knowledge is now added to the equation. Growth accounting has, since Solow, made knowledge and technology *en*dogenous, not only in the eyes of economists but in the minds of policy makers.[9] Why is this so?

Knowledge

Knowledge is largely a public good. Unlike physical resources, it can be used and re-used over time without losing value. Its intellectual property can be transferred locally and internationally without losing ownership. Uncertainty is high in its production (*i.e.* research), but this drops rapidly as it is imitated and diffused. (But so do the returns from lower risk activities.) Of course, firms pursue strategies of being world leaders, close followers, or imitators; each carries with it a variety of risk and investment requirements, particularly across industries. This in itself has significant implications for Scottish firms and regions. Indeed, exactly because there are also considerable spillover effects in terms of location decisions by firms, Scotland is every bit as viable as Singapore. Neo-Ricardian comparative advantage can be overturned by the governance of knowledge and innovation. Neo-Porterian advantage can be overturned by the constructed advantage brought by ingenuity, innovation and knowledge use.

Today, customisation, niche production, knowledge and networks deliver *increasing* returns. The factors of production have changed and the contexts of smaller economies differ. In the cases of the Nordics, Scotland and Canada for

 GLOBAL KNOWLEDGE FLOWS AND ECONOMIC DEVELOPMENT – ISBN 92-64-10766-5 – © OECD 2004

example, proximity to the former Soviet Union, the United States, the United Kingdom or the European Union most assuredly have had an important impact on industrial performance and structure. The cases of Australia, South Africa, Singapore and Taiwan, for example, differ again, but not because of their immediate proximity to large markets but more because of colonial histories. All are small or medium-sized economies, lacking scale or scope in the traditional sense. But all are competitive in the global economy based on innovation and knowledge-intensive activities. Thus the question remains on the front burner of decision makers: "given an uneven playing field in comparative terms and a deficiency of scale and scope, how can we build advantage in the new economy?" This is a question that both Scotland and Canada are asking today.

Constructed advantage

Evidence shows that successful regions understand three levels of issues on which policy makers need to ask a series of ongoing questions. These could be stylised in the most simple of ways.

Q1: *Why are people "here" and why would people come and stay?*

In Canada's eastern Maritimes, the social demographic shows the highest number of universities per capita; the highest educated population and the highest aggregate level of unemployment in the country. This is because, while people are born in Nova Scotia, New Brunswick and Prince Edward Island, they move to Toronto or New York for graduate school and only return to the Maritimes for retirement after a successful career elsewhere. From an economic development and innovation perspective, cultural charm and familial ties are therefore not enough. They *will* come however *if* there is opportunity for business development, risk capital and market access. Quality of life of course *does* matter, but for many "techies" this means: is a place safe, can I run (ski, bike), can I get to work easily (this is now an issue for Londoners), etc. But from a talent/investment/growth point of view, this is subsidiary.

Q2: *Why is investment drawn into a region?*

Responses to this question would include: access to smart people; access to infrastructure; access to leveraged funds; access to new and adaptive technologies, footholds into new and potential markets, (Scotland if it chose to could have a unique window on the European Union), and; a complementary regulatory regime. Scotland has very smart people, but not enough to be known globally as *the* place to go – as *the* gateway. Differentiation is key. Why go to Glasgow when one could go to London or Dublin, getting subsidies, talent, money and access to more than 400 million people? Moreover, branding efforts often ring hollow for investors. Where is "silicon glen" or why say "the tartan tiger'? (Parenthetically, Canada, Wales and Ireland have also used the "Tiger" metaphor despite the fact that no one has ever seen a tiger in any of these locales and even Dylan Thomas lamented about a "Wales without wolves.")[10] No one is looking for a tiger. They're looking to invest in entrepreneurial opportunity, in locations that can grow smart firms, employ smart people, and penetrate world markets. Regional and small national governments often fall into hyperbole. This actually deflects investment, attracting "vulture capital" instead of *venture* capital. Gerhard Mencsh (1978) understood this deeply when he wrote his book on *Stalemate in Technology* in 1978.

Q3: *What should public decision-makers do?*

- Recognise, *deeply*, the nature of innovation. Many do not. Underneath the obvious factors of production, innovation is based on risk, uncertainty, expertise, and networks. Sustainable communities are built on local networks and a spirit of collaboration. Public decision-makers can play a critical role in this.
- Engage local industries, university instructors, higher education leaders, not-for-profit organisations, youth groups (remember that fascinating and viable knowledge based activities are coming out of "cyber-punks" – indeed, mayors in numerous cities across North America are including such "multi-media development software engineers" in policy design) (Florida, 2002).

How can these two basic proposals be conceptualised and achieved? One of course could easily fall into the thoughtful area of Schumpeter or Hayek. But *practically* we can say, "let's look at our region, our future, and our potential as a community." OK. We may wish to draw a diagram such as Figure 5.2.

Figure 5.2. **Ingredients for innovation**

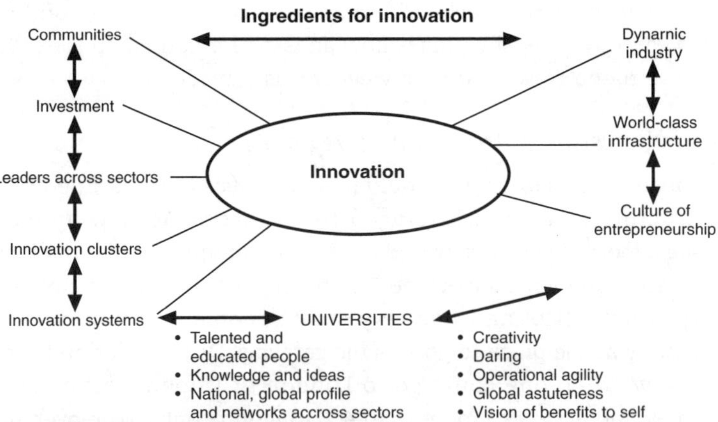

Source: Heather Munroe-Bloom, University of Toronto.

What does this show? Not much, but it does make us alive to the fact that innovation is a "full contact sport". For innovation and growth to occur, a region needs – no, demands, collaboration. Otherwise why stay, why commit, why invest? This simple figure actually *challenges* leaders and decision-makers at every level.

One can see how this has been recently adopted, albeit not perfectly, by the City of Ottawa (Figure 5.3).

Figure 5.3. **Innovation pipeline**

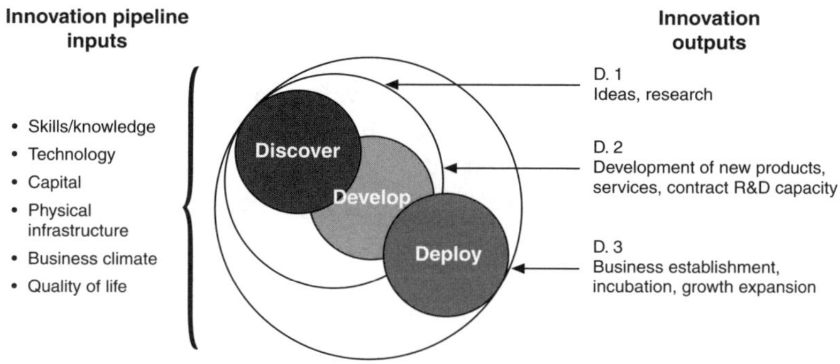

Source: City of Ottawa, 2002.

Or see how other officials have, in Canada, portrayed the "building of a region", the constructing of advantage (Figure 5.4).

Figure 5.4. **Constructing advantage**

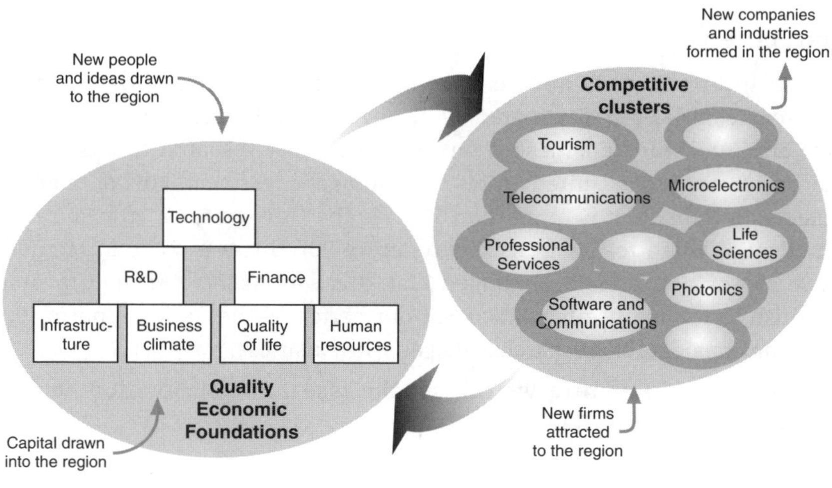

Source: City of Ottawa.

Thus we can draw some issues for Scotland, and point to some lessons from Canadian cities that are trying to develop a learning region.

A learning region needs leadership from all local actors, firms, universities, government, and non-governmental organisations. This requires engagement, vision and debate in order to see your community as viable and vibrant in the global economy.

The innovative community also needs physical infrastructure, including airports, good roads, local amenities and institutions that offer support for investors and skilled people. Investment today is *not* going into New Haven even though it has Yale University and it is close to New York City. This is because the city has not dealt with the crime rate, has not dealt with re-vitalising the downtown core through renovation, and only one airline goes into Tweed International Airport so it is hard to get to. Instead, Newark is getting tremendous investment and inflow of talent because of it's ease of access, proximity to New York (20 minutes), good schools (Rutgers and Princeton are near), and so on.

The learning regions of course must have smart infrastructure – it must be a "wired" community – as well as access to the full range of quality capital. This means everything from "love capital" to loans to venture capital and equity. But this is not just a matter of access, but of culture. Is money available for entrepreneurial activity and risk? How is it secured, and so on? In Canadian cities business owners are being asked if they feel their local bank is on their side. This is an important indicator of the local culture for innovation. Does your community have a propensity to innovate? (If not, what can leadership do?) Does your community collaborate? As we've already seen empirically, Scotland tends not to.

Canadian cities

An interesting element is that in terms of collaboration, Canadian researchers, cities and firms are locally engaged and internationally present. Take one indicator – business sponsored university research through grants and contracts. In the United States the average is less than 10% of operating budgets. In Canada, it is more than 17%. At the University of Ottawa (ranked 8th as a Canadian research university out of 90), this was 27%. At Memorial University of Newfoundland, ranked 21st, this was 20.8%.[11] At the University of British Columbia it rates 22%. Each out performs the national norm, thus revealing considerable research linkages with business.[12]

Looked at differently, we can see the research collaboration between cities in the following table.

GLOBAL KNOWLEDGE FLOWS AND ECONOMIC DEVELOPMENT – ISBN 92-64-10766-5 – © OECD 2004

Table 5.1. **International scientific/technological collaborations between selected cities**

Calgary, Halifax, Ottawa and Vancouver, with US and Scottish cities (1991-2000)

	Calgary	Halifax	Ottawa	Vancouver	No. collabs	Total collabs with Canada	CL[1]	Total publications	% CI
Austin	21	8	89	27	142	378	3,347	16 206	20.7
Durham	31	15	242	276	362	1 020	3 758	21 710	17.3
Edinburgh	18	8	36	51	106	382	10 062	22 212	45.3
Glasgow	19	10	37	15	77	399	9 358	20 885	44.8
Raleigh	9	9	15	46	70	219	2 141	13 077	16.4
San-Jose	5	3	19	18	41	116	1 269	5 811	21.8
Tucson	46	14	64	73	184	768	5 353	20 743	25.8
No. collabs[2]	135	60	484	457					
CI	3 422	2 357	7 326	8 083					
Total publications	10 710	7 901	23 214	23 144					
% CI	32.0	29.8	31.6	34.9					

1. International collaborations.
2. No. Collaborations is not the sum of columns (or lines) because articles written in collaboration may concern several cities.

Source: Observatoire des sciences et des technologies – CIRST.

Clearly, Ottawa and Vancouver both collaborate substantially with Raleigh/Durham. A simple explanation is that both Nortel and JDS Uniphase are present in each, as well as the fact that Duke University, the University of North Carolina, the University of British Columbia, the University of Ottawa, and Carleton University all have adapted their engineering, science, law, and management curricula to understand the telecom, photonics and biotech worlds. We have here evidence of a strong, and growing, distributed system of innovation and a strong flow of knowledge.

At the municipal level, Ottawa has been exemplary in constructing advantage. Ottawa, which was once known as a bureaucrat's town (being the nation's capital), has become overwhelmingly a technology town. Since 1993, the number of advanced technology companies has increased from 465 to nearly 1 400. High tech employment has grown from 30 000 to more than 70 000. Comparative business costs are the lowest against Atlanta, Raleigh/Durham, Washington DC, Boston and San Jose. Ottawa is the 4*th* largest urban area in Canada with a regional population of 1.2 million. Markets involving 150 million people are within 2 hours flying time. Advanced technologies generate $20 billion in annual sales, 90% of which are through exports. And Ottawa agencies – universities, City Hall, firms, and industrial associations such as the Canadian Advanced Technology Alliance (CATA) and the Ottawa Center for Research and Innovation (OCRI) – all work hard constructing advantage and linking stakeholders.

An example of the importance of alliances and constructed advantage can be seen in the on-going exercise that is geared towards stimulating local action around high tech. In recognition of the importance of constructing advantage in an increasingly competitive knowledge-based international economy, the Canadian Advanced Technology Association Alliance (CATA), which represents more than 40 firms, has undertaken a program entitled "TechAction". Organised in conjunction with the city mayors, CATA has – with KPMG and IPSOS-NPD (a survey specialist) – conducted in-depth telephone interviews with approximately 100-150 local Chief Executive Officers, managers of firms and labs, university administrators and researchers, and firm directors. They have also published one page "write in" invitations in local papers for anyone in that community that has a view as to the strengths, weaknesses, character, and future of the community. Facilitated Town Hall meetings are then carried out at City Hall and led by the Mayor with discussions – by the community, for the community – dealing with four key elements. These are leadership, people, money and infrastructure. Some of the aggregated views for cities visited so far are represented in Table 5.2 below. These discussions have led communities to share views as to who should do what in their collective quest to ensure a vibrant community future and in their quest to construct advantage through alliances and shared, deliberate, but flexible strategies. A postulated "perfectly innovative community" is subjectively given the scale of 5 on four axes. In responses, the respondents gave a rating based on "gut feel" about where their community "is". This is represented graphically, giving leaders and participants a sense of where the community thinks it needs strengthening in order to be a fully sustainable and dynamic innovative community.

Table 5.2. **Local views on current strengths of their communities**

	Leadership	People	Technology	Money
Ottawa	3.89	3.6	3.74	3.06
Vancouver	2.99	3.6	3.37	2.69
St. John's	3.5	3.69	3.5	2.79

Note: Scores are on a scale of 1-5, with 5 representing the greatest strengths.
Source: CATA.

Ottawa is an interesting case. For decades, the city's main employer was the federal government. Over time, however, with the critically important presence of the National Research Council, a great number of highly qualified researchers were attracted to Ottawa since the 1950s and a large number of high tech firms have been spun out, created and grown. Examples include what is now Nortel Networks (which started as Northern Electric then

Northern Telecom), Newbridge, JDS Uniphase, Cognos and Corel. Today, high tech is the dominant employer and diversity in the knowledge-based sectors has emerged over the past decade. Alliances have been central to this value-added – high standard of living – transformation. The Ottawa case is not unique, but it is illustrative in surprising ways. It has now achieved a strong reputation, in particular, in telecommunications, photonics and biomedical devices. Direct alliances between the two major universities and colleges and the private sector have changed curricula in engineering, science, management and law. The current and previous Mayors were cheerleaders for high tech. Ottawa now has air links to Chicago, New York, London, Boston, Raleigh and other high tech and financial centres.

Industry associations like CATA are helping to introduce Ottawa firms with counterparts in Tucson Arizona, Austin Texas and Monterrey Mexico. Even the indirect, but important, impact of the local professional hockey team, the Ottawa Senators, has played a role given that when viewers watch hockey across North America, they see the advertising on the rink boards featuring Nortel, JDS, and formerly Newbridge (now part of the French firm Alcatel) and JetForm (now part of the American firm Adobe) and numerous other internationally known tech firms that call Ottawa home. Thus the community strategically and actively uses constructed advantage and community alliances to forge its future.[13] Differences between innovative communities are telling.

In St. John's Newfoundland, hundreds of miles from continental North America's eastern coast, the Mayor was present for the duration of the TechAction Town Hall meeting. The audience was stimulated, and collectively viewed their marine-locked way of life as precious and unique while at the same time saw their futures, collectively, as being firmly in the knowledge-based global economy. The role of Memorial University was seen as key, as were the federal laboratories of the National Research Council. Realising their collective futures in petroleum, cold-ocean drilling, marine biology, fish stock management, as well as in information technologies and geographic positioning systems would – in their view – require community alliances, both within Newfoundland and across Canada and the world. Whilst money and infrastructure were clearly recognised as being weak, yet important, leadership from all sectors was seen as key for alliances success.

One could also note Vancouver on Canada's West Coast. It has a natural deep harbour and access to a beautiful rain forest and mountains. The temptation therefore to rely lazily on traditional neo-Ricardian comparative advantage is palpable. It has two major universities but, as they are both nestled in the mountains, somewhat remote to downtown, there is little interaction or collaboration. Yet it does interact extensively with the US north-west and the trans-border region has given itself the name of Cascadia.

Technology, investment, and talent are beginning to really flow into the region and build a learning community.

Other Canadian cities are successfully following their own path to innovation and economic development. Montreal, for example, closely follows San Francisco, Boston and Seattle in terms of high tech employment per capita, outpacing Chicago, Houston and Atlanta. Its high tech exports have grown from about $6 billion in 1991 to more than $24 billion in 2000. As a city, it has *decided* to become outstanding in biopharma, IT and – because of history with long present firms such as SNC Lavalin and Bombardier – aerospace. In 1999 Montreal firms invested more than $1.5 billion on R&D (double 1990) and – with 4 major universities and roughly 2 100 professors – the output of published scientific papers is comparable to Dallas, Seattle and Atlanta.

Conclusion

In each brief case of Ottawa, St. John, Vancouver and Montreal, the community has deliberately and forcefully developed a new economic vision with innovation at its centre, and each has tried to leverage its local strengths in a high value-added set of activities. They are constructing advantage. In so doing, they are attracting talent, developing talent, drawing on the world pool of foreign direct investment, and competing globally. There are lessons here for Scotland.

Notes

1. The assistance of Professor Yves Gingras of the University of Quebec in Montreal, Tyler Chamberlin of the University of Manchester, Jessica Vinograd of the University of Ottawa, Dr. Mark Freel of the University of Aberdeen, Professor Philip Cooke of the University of Wales at Cardiff and Barry Gander of the Canadian Advanced Technology Alliance (CATA) is gratefully acknowledged. All interpretations, however, are solely the responsibility of the author.
2. *www.scotland.gov.uk/stats/bulletin/00234-00.asp.*
3. In the last research assessment exercise, 6 Scottish departments received the highest "5 star" rating.
4. This number is imprecise as the Knowledge Exchange also notes as an aside that the electronics industry employs 41 000 people but this figure does not jibe with the otherwise stated figures.
5. Scottish Keyfacts compiled by the Knowledge Exchange (Glasgow), 21.11.02.
6. Spectrum International Benchmarking Study, 1999, cited in A Smart, Successful Scotland (2001).
7. GERD is Government Expenditure on R&D.
8. The anti thesis of Timothy Leary's motto, "turn on, tune in, drop out", it would seem.

9. No longer is technology T seen as being equally accessible over time and space, to all "rational actors" who possess "perfect information". See any of the textbooks on macroeconomics by N. Gregory Mankiw.
10. Politicians have recently referred to Canada as the Northern Tiger, Wales as the Welsh Tiger and Ireland as the Celtic Tiger. Even India is having trouble holding on to its tigers!
11. Canadian Association of University Business Offices, Financial Information of Universities and Colleges 2000 2001, Ottawa (*www.caubo.ca*).
12. See *www.researchmoneyinc.com*.
13. *www.ottawaregion.com*.

References

Florida, R. (2002), "Bohemia Index", *The Rise of the Creative Class*, Basic: New York.

Guehenno, J.M. (1995), *The End of the Nation State*, University of Minnesota Press: Minneapolis.

Mencsh, G. (1978), *Stalemate in Technology*, Ballinger: New York.

de la Mothe, J. (ed.) (2002), *Science, Technology and Governance*, Continuum: London.

Nimijean, R. (1998), St. Johns New Brunswick as an Emerging Local System of Innovation, in John de la Mothe and Gilles Paquet (eds) *Local and Regional Systems of Innovation*, Boston: Kluwer.

Ohmae, K. (2000), *The Invisible Continent*, Harper Business Press: New York.

Sassen, S. (2000), *Cities in a World Economy*, Pine Forge (Sage): Beverly Hills.

ISBN 92-64-10766-5
Global Knowledge Flows and Economic Development

Chapter 6

Internationalising knowledge flows: Pacific experience

by

Paul Frater (Industry New Zealand, New Zealand)

The strategies of Pacific countries and regions for connecting to global knowledge flows are examined. The chapter underlines the importance of links to export networks, angel investor networks, industry clusters and research consortia for the internationalisation of innovative small firms. Examples of successful connection to global knowledge flows are taken from Silicon Valley, Korea, Taiwan and Singapore. A number of New Zealand initiatives are also outlined, including the commercialisation activities of the Crown Research Institutes, the wood processing strategy, research and technology networks, the education and skills strategy, immigration procedures and investment funds. An eight-step programme is outlined for building high growth, high value added industries.

Internationalising knowledge flows: a Pacific perspective

Key vehicles for internationalisation

Scotland has set an objective of returning to the top quartile of the OECD rankings in terms of income per capita. New Zealand (currently somewhat lower down the scale) has a goal of returning to the top half of the OECD rankings, with sustainable economic development. Both goals imply a substantial lift in productivity for the respective nations, and neither can rely upon their domestic markets to generate the necessary growth opportunities, given their small populations of 5.5 and 4.0 million people respectively. Therefore, engagement in trade with the global market place is essential for both, especially New Zealand, which lies outside the major powerhouse regions of North America, Europe and North Asia.

Within this context, locally headquartered international business activities are the key vehicle for building the commercial activity that can deliver the desired scale of growth in productivity. Such a mechanism is not the only strategy available. For example, Israel creates many high technology start-ups (driven in part off its high defence research and technology expenditures) but these firms are usually sold to US firms for global expansion, and are typically transferred to the US as part of that process.

Locally headquartered in this context can mean an operating or regional group of a multi-national corporation (MNC), or the headquarters of a locally owned international business. In small economies, niche market strategies are an immediate corollary, but the Internet is creating opportunities for new global business models that mean that physical isolation can be overcome, and good international business can now be run well from these locations.

If growth in productivity is the goal, then a nation's competition in global markets cannot be based upon "low-cost" strategies – instead it must be based upon innovations which are sufficient to create a market niche which can be dominated and developed. It is also possible that a global market sector may be created as Scandinavia has achieved in wireless communications. In order to create a significant difference to per capita prosperity, the innovations must be sufficient to create an internationally recognised activity. In Silicon Valley, this is interpreted as being global sales of USD 1 billion or more. In Scotland and New Zealand, niche markets of this size are large comparative to the domestic market,

 GLOBAL KNOWLEDGE FLOWS AND ECONOMIC DEVELOPMENT – ISBN 92-64-10766-5 – © OECD 2004

and smaller niches are also available, and may even offer better value adding opportunities.

National and regional innovation systems

The national and regional innovation system therefore becomes very important in this context. The high growth performance of regional innovation systems such as Silicon Valley California, Cambridge in the UK, Hsinchu in Taiwan, and the "third Italy" all offer different system models and dynamics. In this context, we wish to explore three dimensions of innovation systems.

1. Research-led or science-based innovation systems.
2. Practice based innovation systems, where enterprises generate new innovations and business models from all parts of their organisations, not just from research or product design and development departments.
3. Technology acquisition, diffusion, adaptation and learning systems such as those pioneered by Japan and the "Asian Tigers" – Korea, Taiwan, Singapore and other South East Asian nations.

Earlier chapters by Phillip Cooke, John de la Mothe and Bjorn Asheim have covered the research-led innovation system in considerable detail, including matters such as entrepreneurial regional innovation systems, the transfer of research knowledge into enterprises and commercial markets and the strength given to these activities through regional clustering. Those papers have also covered issues of biotechnology, which is of increasing importance to both Scotland and New Zealand, and networking, distributed innovation systems and knowledge workers. In particular, de la Mothe addresses the issue of cross-national border inter-regional collaboration.

Here, we want to view these research-led innovation systems in terms of their outcomes. The results of the Stanford Institute for Economic and Policy Research (SIEPR) research project[1] on Silicon Valley and similar innovation systems is particularly instructive. The project investigated clusters in Ireland, Cambridge UK, Israel, Scandinavia, India, Taiwan and within the US, Northern Virginia and the Silicon Valley of the 1960's. All except Cambridge have achieved double digit growth figures. All, except Scandinavia, were active in the same vertical ICT market. That market was dominated by Silicon Valley, and these new clusters established centres of technology, through participating in a worldwide production network which was essentially an extension of Silicon Valley. They are complementary players to the Valley's production systems.

In Scandinavia, wireless communication technologies were developed into a leading global industry, based upon the European market and its adoption of a region-wide communications standard – GSM. A new vertical market was created and Scandinavian and European firms became the dominant market players, extending their standards and systems to other world markets, especially Asia. These models do not apply only to fast-growing high technology markets. New Zealand's experience with similar issues in its wood processing industries will be discussed below.

Issues of small-scale and peripherality

New Zealand research system reforms

New Zealand has a long legacy of state science research institutions of world-class standards. Since 1984 it has been pursuing a programme of improving its innovation performance with increasing intensity and urgency. Until the new millennium, it had little to show for its endeavours outside its traditional industries of year-round pastoral grazing, horticulture and primary processing industries.

Major reforms have included:

- 1992: the consolidation and incorporation of all crown research laboratories into crown-owned corporate entities (Crown Research Institutes, or CRI's) within a contestable funding system.
- 1996: Major extension of the boundary of research activity towards commercialisable research activity.
- 2000: formation of Industry New Zealand as a regional and industrial economic capability-building agency.
- 2002: a major government focus on a "Growth and Innovation Framework", and expanded resourcing of its innovation and economic growth activities, including the development of new public/private funded Venture Capital funds, drawing on accumulated reserves of the CRIs to provide initial public sector investments. An Angel Fund network was established.

Only in 2001 did New Zealand's economic growth rate begin to rise above that of the OECD average, and some of that was attributable to the nation's emergence from a 3-year drought. Despite consistent attention to the issue of growth and innovation, New Zealand was a long-running poor performer.

It would appear that improving the innovation system alone has been insufficient. Indeed, the whole process has appeared to be bedevilled by the sale to overseas companies in the same industry of a series of rising new Kiwi commercial stars. Many have subsequently ceased operations in New Zealand, and transferred the activity overseas. New Zealanders have found this to be a very frustrating experience, especially in the light of the nation's limited capability to enter into similar activities overseas.

However the CRIs have continued to develop their commercialisation agenda:

- The "spinout" strategy grows in strength across the system, as the institutes establish new businesses, encourage staff commercialisation buy-outs and

 GLOBAL KNOWLEDGE FLOWS AND ECONOMIC DEVELOPMENT – ISBN 92-64-10766-5 – © OECD 2004

joint ventures with local and international partners. The Vote Research, Science and Technology (Vote RS&T) system reinforces these trends by employing private sector engagement criteria in its funding, and seeking evidence of successful transfer of research intellectual property into the private sector.

- The CRIs have ceased to be inwardly focussed, self-sufficient/absorbed research institutions. They actively seek international research and industry liaisons and collaborations to build necessary "connectedness" to give their work international scrutiny and opportunities to participate in the building of global industries and value chains. Building, developing and gaining commercial momentum from these networks is the brief of an expanding team of Commercial Managers in the CRIs.
- Research collaborations between universities, CRIs and private enterprise are supported by a Research Consortia programme. The wisdom of this has been widely accepted, and the process voluntarily extended by the various institutions. Three aspects have been important drivers of these initiatives:
  - The need to engage more private sector businesses in formal R&D work.
  - The need to design R&D programmes that the private sector would be prepared to invest in.
  - The realisation that a small economy cannot afford any significant duplication of key facilities in a number of stand-alone laboratories/ research projects – sharing and collaboration is essential.
- The Research Consortia have enrolled medium-sized firms into R&D projects, which represents a significant advance on the previous situation, where only large scale enterprises (LSEs) were active in R&D activities with the CRIs.

Despite the evident uplift in private sector involvement in formal R&D, and the increasing dynamic in CRI internationalisation and intellectual property commercialisation processes, doubt remains as to whether New Zealand has achieved a genuine upgrading of its innovation and productivity performance – or if it has only created a period of local early stage development, which will then return to the pattern of sale of small and medium scale, internationalising companies to overseas companies, and their operations being transferred in whole or large part, offshore. Standing well outside the major trading blocks of Asia, North America and Europe, New Zealand faces a perpetual challenge of the periphery – being a branch office economy, and often a sales office branch at that.

The improved integration of the national innovation system may simply have extended the period of New Zealand ownership of innovative companies but not changed the long-term probability of overseas sale and ultimate transfer offshore of the enterprise and the technical and managerial personnel. If these concerns are correct, then New Zealand will be extracting much more value from its knowledge generation, but will still not be securing the economic development potential it could.

Micro enterprise economies

Related to this challenge may be the micro and mini-scale of the New Zealand population of enterprises – see Table 6.1. More than 91% of New Zealand's enterprises are micro, employing less than 10 people. LSEs are not statistically significant. This contrasts sharply with countries such as the US and Korea, where micro enterprises represent less than 45% of the business population, and with Japan where they represent only 53%. In the United Kingdom the proportion of micro enterprises is 72%. It was reported that the Scottish population would have a higher proportion of micro enterprises than the UK on average. With only 1.1% of enterprises employing more than 50 employees New Zealand CRIs have a very small potential customer base in their domestic markets.

Table 6.1. **Distribution of establishments by numbers employed, selected countries**

		Employment size class					
		0-9	10-49	50-99	100-499	500+	Total
		Percentages					
New Zealand	2001	91.3	7.6	0.7	0.4	0.0	100.0
Finland	1996	84.5	11.2	1.9	1.9	0.5	100.0
Italy	1995	81.6	15.9	1.4	0.9	0.1	100.0
United Kingdom	1996	71.5	20.9	3.4	3.5	0.7	100.0
Sweden	1996	71.3	21.2	3.6	3.1	0.8	100.0
Japan	1997	53.2	37.2	5.3	3.8	0.5	100.0
United States	1996	44.6	28.0	6.2	7.7	13.5	100.0
Korea	1995	44.3	46.1	5.5	3.6	0.5	100.0

Source: Statistics New Zealand.

Earlier research into a comparison of manufacturing enterprise formation and average size in New Zealand, Ohio and Sweden (Carlsson, 1996) had identified these differences. While total manufacturing employment in all 3 states had declined by 20% over the period 1976-1993, and all had experienced markedly different macro-economic environments:

- The number of manufacturing establishments:
  - remained constant in Sweden;
  - declined by 9% in Ohio;
  - doubled in New Zealand.

- The average establishment size therefore:
  - declined by 26% in Sweden to 27.3;
  - declined by 20% in Ohio to 56.5;
  - reduced by two thirds in New Zealand to 11.5.

Thus Sweden's average manufacturing enterprise could be classified as "small", Ohio's as "medium", but New Zealand's as "mini" and fast approaching "micro".

While issues of market size/access, industrial legacy and current industrial mix are all influencing these results, there is a clear impact arising from the isolation and peripheral location of New Zealand industrial activity in global industries and value chains. Add to this the dominance of micro (less than 10 employees) and mini enterprises in the New Zealand industrial structure, and the impact upon its national innovation system is profound (discussed below). Scottish Enterprise reported that it faced similar (but not as extreme) small scale and peripheral issues to New Zealand. Countries with very high populations of micro and mini-scale enterprises may require more than simply good design of a traditional research-led innovation system.

New Zealand system review

In a review of the performance of the new and fast-growth businesses of the New Zealand economy with Investment New Zealand's North American Advisory Board in early 2003, the following were noted:

- The New Zealand growth companies operating in the international market place are taking too long (10-15 years) to reach a size and level of engagement where they have recognition and credibility. This compares with targets of "within 5 years" typically used in North American venture and development capital circles.
- The business models adopted by new technology spinouts and new initiatives of New Zealand firms are managed in an incremental innovation strategy, rather than the context of radical innovation as Gary Hamel defines it.
- The business plans being submitted by new technology and fast-growing businesses are limited in terms of their appreciation of and strategies around:
  - International competitors.
  - The business models and value-chain structures of the international markets they are engaged in.

- The venture and development capital strategies for investors imitate those of North America (aiming at an IPO (Initial Public Offering)) when the reality of New Zealand's exit experience as noted above, is that most exits are made through "trade sales" to overseas companies.

Implications of micro-scale structure

New Zealand's enterprises are predominantly micro or mini, not even small. Their domestic market is small. Thus, their challenge in scaling up to internationally relevant and robust stature and capability is some orders of magnitude greater than that for counterparts in medium and large market economies. The domestic capital market is small, and has not attracted sufficient international investor involvement to provide a viable basis for IPO investor exit strategies.

In part, these issues may explain why New Zealand fast-growing enterprises engaged in international market-building have so frequently been sold to overseas counterparts:

- Because of the extra degrees of scale that have to be built up from the domestic base to become an international market leader, the New Zealand firms are likely to be several years behind in development, other things being equal, and so have less financial capacity among other things.
- Options for domestically financed investor exit strategies are severely restricted.

Overseas transfers of operations developed in New Zealand have often followed trade sales of the equity of New Zealand businesses creating a migration of top management and technical talent. These are the very resources that are the vehicle for lifting productivity and per capita incomes. The challenge is to break this mould. A deep investigation of the overall design of the total innovation and internationalisation system is required.

The recently published study of New Zealand business practices and performance (Knuckey and Johnston, 2002) notes the concern over the low level of formal R&D in New Zealand firms, but argues that other sources of innovation are also important, and may be more so in a small open economy where a lot of the necessary technology can be purchased overseas if benefits of scale and specialisation apply. In these circumstances, innovation capabilities within firms, skill levels and an innovation supporting institutional framework may be the critical requirements.

The Firm Foundations study found that a reasonable proportion of New Zealand businesses were investing to build their innovative capabilities, and many were purchasing machinery and equipment as an important vehicle for this. Almost half were investing in employee skills to support new innovations in products, processes and services. Yet only 20% indicate that they regularly undertake R&D. At the same time design processes are emerging as a major response of New Zealand firms to customer feedback and competitor analysis.

New Zealand business internationalisation process

The New Zealand conditions have thrown up an economic development experience which is counter-factual to that which models the experience of the larger economies. A research project, Competitive Advantage New Zealand (CANZ) has been investigating the experience of the New Zealand-owned internationalisation success stories. CANZ has investigated the development of innovative manufacturing and service organisations and their building of sustainable competitive advantage within the global economy (Campbell-Hunt, 2001).

In brief, micro and very small enterprises generate innovative breakthroughs that have international market potential from within their own firms and customer networks. But the internationalisation challenge for a small firm building from a small and isolated market is huge. The SMEs have to enter global markets when they are smaller and more immature than would be the case in most other developed countries. The overseas market growth is explosive – these firms can expand up to ten-fold in 3-4 years. This growth, colloquially labelled "the gusher", places enormous strain on financial and managerial resources.

CANZ has investigated the evolution of competitive advantage in 24 exemplary New Zealand firms. All achieved and sustained competitive strength in world markets, and in so doing, have been vehicles for the internationalisation of New Zealand knowledge. These firms undergo a series of transitions that can be summarised as:

1. The development of a successful innovation (which includes the successful realisation of knowledge in the form of goods and/or services).
2. Initial engagement with international markets in a "sow and reap" strategy which inexpensively gains an appreciation of market potential and the need for design and other modifications to meet market needs and competition.
3. The lifting (or often continuation) of high-level R&D activity, typically at around 10% of sales, and the focus of the enterprise on a promising product or market.
4. The "gusher experience", typically flowing from a second generation of the product or service.
5. Keeping the enterprise "in balance" as its marketing, production, logistics, training and management systems are rapidly expanded.

6. Completion of the transition to a "future capable company". Often the innovation comes from a small enterprise, operating in low-margin market segments. They are at the bottom of the value curve. The transition to higher value/higher margin points on the value curve is essential to provide the equity returns necessary for growth, but this transition is not automatic.

As the economy's population of sustained internationally competitive businesses builds, so too does the nation's industrial absorptive capacity to partner in the development of new knowledge and technologies developing in its academic and public research laboratories. Similarly, the absorptive capacity for spinouts from the business sector is also increased.

The average CANZ exemplar firm employs around 200 people. It earns 75% or more of its revenues offshore. These are medium scale enterprises, with some LSEs having been established. In each case, an international partner or advisor has been sought out to supplement the firm's strategic resources. Along with key international customers, these relationships have assisted in such tasks as the development and tightening of the product range, market positioning, refining market propositions, technology acquisition and development, and material and components supply. In short, the companies have required key international market/technology partners in their industry, and through them have become engaged as part of the industry leadership – active in the shaping of the industry structure.

Interestingly, the New Zealand companies have generally tended *not* to locate manufacturing or service development operations offshore. Thus, despite their explosive gusher growth, they are still accruing benefits from economies of scale. Some other high growth New Zealand companies have relocated manufacturing operations to Australia and to China, but this is somewhat atypical of the general New Zealand experience.

Forming domestic and overseas market relationships appears to be a key attribute. Kiwi culture, with its informality and openness to communication across all strata of society provides New Zealand firms with a distinctive advantage in this respect (it may even be a reflection of the isolation and small size of the country). As a result, the firms were able to form rich networks of relationships, and to apply the learning that flows through them. The companies leveraged that learning within their organisations – technologically, geographically, financially and reputationally.

Business networks and clusters

Export networks

Some micro, mini and small enterprises have outstanding knowledge, experience and business models. But in small economies, gaining critical

GLOBAL KNOWLEDGE FLOWS AND ECONOMIC DEVELOPMENT – ISBN 92-64-10766-5 – © OECD 2004

mass and significant capability and capacity requires an early move into international markets. Experience at Trade New Zealand suggests that few micro-small enterprises can invest time and other resources to make a sustained commitment to international market development. Their networks are weak. Often their production capacity is too small to deliver the volumes required from market success. But these requirements can be achieved through the creation of export networks with other micro-small enterprises, thereby giving them market reach, capacity to meet export orders and the (shared) resources to undertake market research and development. Trade New Zealand's export network programme is heavily over-subscribed.

Angel networks

In part, these outcomes also reflect a relatively weak venture financing system, and weak business planning on the part of New Zealand SMEs. The initial global market engagements seem not to be sufficiently well planned and alliances not made. Angel investor networks have a major role to play in this regard, and have become increasingly active in New Zealand over the last 5 years.

Clusters

Industry New Zealand has programmes for the development of industry clusters. Not only are these seen to be critical for the deepening and broadening of local production systems, they are also expected to:

- Generate opportunities for the development of viable specialist firms and activities that would otherwise not be possible if the micro-small enterprises remained in "maximum self-sufficiency" mode.
- Create the necessary critical mass to enable business networks from within the cluster to enter international markets professionally – in other words to give the business international market reach. Otherwise many micro-small enterprises could never realise their international market potential.

Research consortia

Networking strategies have also recently been introduced in the Vote RS&T system through Research Consortia. This concept was developed from Australian initiatives. In the New Zealand model, extra funds were made available in the Vote RS&T for the Consortia, but the funds had to be matched by private sector subscriptions as well. In the event, the public sector funds have sometimes been matched several fold.

The research consortia typically have a number of tertiary education and research institutes collaborating with a group of private sector companies. The private sector counterparts include large and medium scale enterprises –

and some specialist micro-small scale enterprises. The consortia include both domestic companies and the local operations of overseas MNCs. Most recently overseas branches of overseas- headquartered MNCs have become involved in New Zealand consortia.

The involvement of overseas MNCs is seen as important in two dimensions:

- They provide access to research, technology and market networks that otherwise would be very difficult for New Zealand to access.
- They offer extensive international market and customer support networks for intellectual property commercialisation for the joint ventures they participate in from within the consortium.

The joint ventures are seen to be critical to both accelerating the international market development of New Zealand intellectual property, and of New Zealand companies. They help to anchor key support, development and business activity in New Zealand, because of the local ownership element, and the further research activities of the consortia.

Collectively, these four networking strategies – export networks, angel networks, clusters and research consortia – have the potential to both lift the proportion of New Zealand enterprises active in international markets, and the proportion active in formal R&D. The relationships with overseas MNCs should contribute to an acceleration in the pace of internationalisation of New Zealand knowledge and intellectual property.

Securing high growth opportunities

Silicon Valley

The SIEPR research project on Silicon Valley and its imitators asks the same questions as CANZ, but from a regional perspective – what forces create new clusters of entrepreneurship-led growth? Each region examined developed ICT clusters where double-digit per annum growth was achieved in:

- The number of new enterprises.
- ICT revenue.
- ICT employment.
- ICT exports.

The study noted the important effect of the ability of the clusters to capture both direct and indirect external effects, including:

- Rapid market technology development diffusion.
- Ability to attract more specialised inputs for the production system including:
  - technical personnel;

- research laboratories and projects;
- venture capital;
- new business/entrepreneurship support services.

The result was that innovations within these local production systems were more highly levered, and more quickly executed, lifting the return to invention. Commercialisation of knowledge is pushed "faster and closer to markets". A second key result is that with external success, a virtuous cycle of positive feedback was established, and returns to entrepreneurs in certain instances, venture capitalists, technologists, and those in key supplier industries became enormous. The capture of rents by producers and the regions in which they are located and headquartered is the opportunity which New Zealand and Scotland seek in their bid to recover their places in the OECD per capita income rankings.

The development of Silicon Valley has been extensively analysed and written up, most recently by Lee, Miller, Hancock and Rowen (2000) and the May 2002 OECD delegation to Silicon Valley (Chapple, 2002). The Valley's early foundations go back to the formation of Hewlett-Packard in 1939. The development of the Valley economy since the Second World War is very instructive for the knowledge flow and industry internationalisation issues being considered here. Briefly:

1. Research Access: Through events such as the return of William Shockley to his home town of Palo Alto, the Valley gained access to world leading research – silicon electronics from Bell on the US East Coast (telephone hubs) and silicon diodes and transistors (first produced by firms in Texas and Southern California).
2. Commercialisation: The process of spin-out was established when 8 young leaders left from Shockley Semiconductor to form Fairchild Semiconductors.
3. Venture Capital: Utilising a New York investment bank Hayden, Stone and Company (Lecuyer, 2000) to secure an investor for Fairchild Semiconductor, the Valley then built its own venture capital industry (on the success of the investment).
4. Industry Connectedness: The partnering of Fairchild with a Long Island based medium sized military contractor Fairchild Camera and Investment gave access to the critical defence industry market.
5. Growth Management: Fairchild hired experienced marketing and commercial management and local electronic technicians to create an integrated research, design, development and manufacturing capability.
6. New Business Model: A new business model was developed by Fairchild that broke the established defence construction industry's norms.

7. Industry R&D: Fairchild (and its spin-outs and competitors) pursued its own technology research and development.
8. An Entrepreneurial, Education Sector: this facilitated a pattern of constant movement back and forth between industry and university. Elaborate social networks focus academic research on practical problems, while infusing industrial activity with up-to-date science.
9. Networked Business Structure: Silicon Valley is horizontally differentiated, with many competing, often highly specialised firms. This contrasts with the vertical integration of the traditional computer industry of the US East Coast (Rowen, 2000 and Saxenian, 1996). With this horizontal structure came the opportunity for distributed computing, "co-invention", and the emergence of competition – between the horizontal segments of the computing industry as well as within the segments.
10. A Regional Technology Loyalty: The Valley's engineers have developed loyalties to the region, each other and to advancing technology (see Saxenian, *op. cit.*) rather than to individual firms or even industries. This creates a highly mobile, risk-taking workforce, and the ready transmission of knowledge through the Valley's companies.
11. Entrepreneurship: The Valley's advantage is its business strategy – the easy facility with entrepreneurship, new firm formation and building business models.

The business dynamic described here, with its horizontal, highly competitive industry structures, spin-out practices and strong commitment to entrepreneurship and global business models is strongly resonant of the Italian industrial districts system.

Lessons from other ICT regional successes

The SIEPR research study concludes that to grow a new cluster, a region has to do two things:

- Develop a new innovation or technology that must then be linked with inventions and advances. They have to be complementary to existing technologies, which in the case of ICT are mostly sold by US-based firms or US-linked multi-nationals. The new region's opportunity is the large and fast-growing demand for the leading technologies.
- Carefully establish the foundations needed for an innovation cluster's take off, including firm building, and skills development to provide plenty of technology workers. This is a medium to long-term investment. Dynamic investment capital networks are vital.

It should be noted that there is risk inherent within these knowledge-based innovation strategies. New technology strategies have to be pursued before their potential can be assessed – because they have to move before it is clear to everyone else.

The patterns that SIEPR identified were:

1. Highly Skilled Technical Labour

   High skilled labour was a pre-condition for high-growth ICT-based entrepreneurial clusters. Clearly, universities like Stanford and Berkeley are widely recognised, but large firms such as Hewlett-Packard and IBM provide significant training in areas such as technical competencies, management and access to industry management networks. Inwards migration and Diaspora returnees can also be important sources of skilled people.

2. Managerial Leaders

   The contribution of returning expatriates underlines the importance of managerial as well as technical skills. Expatriates also bring with them access to technology, market and managerial networks in other clusters where they have worked.

3. New Firm Formation and Firm Building

   Entrepreneurship – Establishing new firms and building the capabilities of firms are important elements in the building of clusters. But some firms must emerge from the start-ups and become leaders of international business activities. They have to achieve recognition and stature.

4. Connections to Markets

   Connections to markets and to the source of demand is critical to the successful commercialisation of knowledge. The SIEPR study identified two distinct patterns:

   - One used by emerging countries where the linkage was founded upon relationships with the main market, and the development of complementary products to those of the existing leading technologies;
   - The second was to position the companies in areas not covered by the existing US leader firms. Some Scandinavian firms, especially Ericsson and Nokia, have successfully employed this strategy to build dominance in wireless hardware for ICT.

5. Cross-Regional Co-operation and Competition

   Linkages with the dominant technology centres of the Western US have been critical for the "catch-up" regions, especially in their growth phase. Cooperation with existing firms seems to be the best way to secure niches that allow complementary products to be developed. SIEPR conclude that more imitative strategies are unlikely to be effective.

   Cambridge in the UK is an example. Its products were similar to those of Silicon Valley, but the Valley had first mover advantage. Cambridge could only cover spaces in the global market not already exploited by the leader. Thus, Cambridge's ICT industries never matched the growth record of the Western US, or of its Asian partners.

The emerging economies adopted "catch-up" strategies (see below), with their diasporas as particularly valuable assets, not so much for technology transfer as for organisational models and industry and market networks at managerial level.[2] In global ICT industry terms, these regions were peripheral. They benefited from the rapid growth of the ICT market, and the ability to exploit new niches and opportunities that emerged without having to incur the development costs.

The Asian Tigers

Following the success of Japan, a succession of East Asian countries have been able to create a significant global presence often starting from an infrastructure and knowledge base inferior to that of the UK and Scotland. For example:

- Japan: NEC, Fujitsu, Mitsubishi and Hitachi.
- Korea: Samsung, Hyundai and LG Semicon.
- Taiwan: Tatung, Acei, TSMC, UMC Macronix and Winbond.
- Singapore: Chartered Semi-conductors Manufacturing.

As one of the most prosperous regions in the US, the San Francisco Bay area has become a focus for migration, both within the US and from overseas, especially from Asia. It has one of the highest concentrations of immigrants of any US region (almost rivalling New York in 1900). Today immigrant workers, mainly Asian, constitute one third of Silicon Valley's workforce.

Because of the professional isolation of the early Asian migrants in the late 1950's and 1960's many started their own businesses, often raising venture capital supplied on the condition that they hire a non-Asian as president of the company. Often this was done after being passed over for promotion in Valley companies.

The Asian immigrants also established social networks, initially to celebrate holidays and family events with others who share language, culture and background. In time professional and business associations evolved from these social networks. Today these organisations are among the most vibrant and professionally active in the Valley – facilitating professional networking, information exchange and mentoring. They have also helped build a two-way bridge between Silicon Valley and the immigrants' home countries.

Saxenian notes that a scarce resource in the complementary industry process is the ability to be able to locate foreign partners quickly and to manage complex business relationships across cultural and language barriers.

The Valley's Asian immigrants were well positioned to play a pivotal role in this environment – creating the social structures whereby even the smallest producers can locate and maintain mutually beneficial collaborations across the Pacific in terms of capital, manufacturing capability, skills and markets.

The success of the Asian ICT regions has been made possible by the nature of the economic and industrial system in Silicon Valley as described above. John Matthews and Dong-Sung Cho (2000) have documented the strategies used by the East Asian nations.

The catalyst for these developments came from the heavy demand for unskilled assembly operators, and the competitive pressures that the newly successful Japanese companies were putting upon the established US and European semiconductor and electronics industries. A major response from companies such as Motorola, Texas Instruments, Hewlett Packard, IBM, Philips and Fairchild was to invest in low-cost assembly operations abroad.

Korea. In Korea, Samsung became the leading global chip producer in 1993, only a decade after launching its first memory chip product. Samsung and the other Korean firms entered the semiconductor industry through their own internal efforts, and by leveraging the most advanced product and process technologies available in Japan, Europe and the US.

Korea assiduously matched this investment with a large investment in technical training and by setting up technical and engineering institutes, which built Korea's "absorptive capacity" for advanced technologies. In 1969, the government passed legislation, modelled on Japan's earlier 1957 version, to attract domestic investment into the electronics and semiconductor industries.

Global industry connectedness was achieved through diaspora and venture financing strategies. Many Silicon Valley semiconductor firms employed US-trained Korean-American engineers. These people were targeted to provide the scientific and engineering leaders that Korea needed to build its own semiconductor businesses. Silicon Valley also had many cash-starved start-up firms, and a horizontally organised and competitive industry. Korean firms were able to offer Valley chip design firms good terms for the manufacture of their chips – but in return the Koreans required the right to licence their designs. So Korea began to acquire critical technology licenses. But the US firms also used Korean manufacturing capacity as alternatives to Japanese DRAM manufacturers.

The Heriot-Watt University Institute of Technology Management visited Korea in 1997. In their report they note that "the identification of new technology in product and process design, and the acquisition and development of that technology is the cornerstone of Korean companies strategic planning". For example, Samsung Electronics R&D is not solely locally based – it is globally organised to capitalise on local expertise in different regions in the world and to be close to markets. Thus, it has an R&D centre in the UK for Global Positioning System technologies, in India for software, Russia for mathematical modelling and in Silicon Valley for semiconductor technologies.

Taiwan. The Taiwanese semiconductor industry was created in the 1980's through a network based on Princeton University, and a widely supported high technology economic development strategy. Again the Chinese diaspora provided a key asset. The Hsinchu Science-based technology park and the Industrial Technology Research Institute (ITRI) acted as key conduits for technology acquisition and diffusion.

In 1989, the Monte Jade Science and Technology Association was formed to promote business co-operation, investment and technology transfer between Chinese engineers in the Bay Area and Taiwan. The organisation is private, but co-operates with the Taiwanese government.

Today, Taiwan is the world's largest producer of notebook computers and a range of PC components such as motherboards, monitors, scanners and power supplies in keyboards. It has manufacturers producing and marketing their own brands. It has semiconductor and integrated circuit manufacturing capabilities that are the equivalent of leading Japanese and US producers. And like Silicon Valley, it has flexible and efficient networks of specialised SMEs that coordinate this sophisticated industry.

A key step in the formation of the Taiwanese industry was the building of a contract manufacturing DRAM facility, manufacturing chips for small Taiwanese and international clients. Government supported the Taiwan Semiconductor Manufacturing Corporation (TSMC) project but insisted on having (majority) private partners. Philips became the leading equity holder, and transferred its Very Large Systems Integration (VLSI) technologies and licenses to it. TSMC quickly developed its own technological competencies, and successfully established itself as a sophisticated manufacturing and marketing operation. Many new, small integrated circuit (IC) design houses were formed in the Hsinchu region.

Singapore. In Singapore, a semiconductor industry has been created through US, European and Japanese MNC investment, and later from other East Asian countries. Through raising its local skill and technological levels from these MNC firms, Singapore now has a cluster of SME service firms, and indigenous wafer fabricators including Chartered Semiconductor, a state-owned company.

Chartered's formation in 1988 was a new attempt at creating an indigenous Singapore electronics industry. Singapore also deepened the technological and operational activities of some of its long-standing MNC partners such as Hewlett-Packard and SGS Thompson, and encouraged closer supply relations with local firms through its "Local Industry Upgrading Programme".

Early on, Singapore liberalised its immigration and work permit systems for foreign professionals (engineers) being employed by foreign firms. This greatly eased a critical issue for MNC operations, and at the same time addressed a key shortfall for Singapore.

Singapore is mounting a major internationalisation drive by establishing jointly-owned technology parks in a number of countries, including China, India, Indonesia and Vietnam. For example, the Economic Development Board has invested in the China-Singapore Suzhou Industrial Park, and will co-invest with firms establishing in those parks (*e.g.* Hitachi Semiconductor).

High-growth SMEs

The OECD has just published the results of a set of comparative case studies of high growth SMEs in eight countries (OECD, 2002). It also shows that exports are a precursor to growth. Exporting *preceded* a period of strong growth in the 7 European countries and the Canadian province of Quebec that were surveyed. The SMEs built their growth on strong local markets and overseas marketing subsidiaries.

Businesses that experienced strong growth had deliberate innovation strategies, developed new products, entered new markets, and expanded their geographic market area. The survey found a much higher incidence of R&D in the high growth SMEs – 33% compared with 13% in the total permanent population, and 70% of the high growth SMEs were engaged in some form of research activity, either directly or indirectly.

High-growth SMEs are innovative organisationally. They undertake process innovation, business reorganisation and marketing innovation. Design activities are of growing importance. Only one-third of the high-growth SMEs operated in markets enjoying natural growth. Growth is primarily achieved through winning new customers, supplemented by a broadening of the product range for existing customers. Direct marketing to customers was a feature of 90% of the high-growth firms. Two thirds depended on a few (5) leading customers and they had a high market share. Partnerships with customers were seen by management to be a key to achieving high growth.

The high growth SMEs attach great importance to staff training and to passing on the benefits of growth to their staff. Nearly 80% offered incentive packages.

There are a lot of resonances with the CANZ results in this study, possibly reflecting the SME focus. The roles of the domestic market and exports are radically different however, the OECD respondents are clearly larger than the New Zealand SMEs, particularly at the beginning of their growth evolution.

An internationalisation process

There is surprising agreement across the studies and experience reviewed above that there is a process that can be used for building high-growth, high value added industries for "catch-up" regions. There are strong synergies with the way that Silicon Valley had to create its regional ICT clustered industry in the late 1950's and the 1960's. The steps can be summarised as follows:

1. Knowledge generation, which may occur within:
   - the public research system,
   - research and development activities in private firms,
   - critical review of business models and activities at all levels of private firms and organisations.

   In the case of catch-up entry into new industries and technologies this activity will be preceded by a process of technology acquisition and diffusion, and the preparation of necessary foundations such as a skills base, a research base and financial capital resources.
2. The seeding of that knowledge into an organisation or team that is competent to develop it. The task here is to create an innovation or service dimension that creates the opportunity to enter international markets.
3. The building of an entrepreneurial vehicle to champion the innovation. This may be a spinout, a new business unit or team within a firm, or a joint venture.
4. The innovation is then exposed to the global market and customer reactions assessed, and competitor responses identified.
5. Typically, another generation of the product or service has to be developed to focus on the market opportunities that have been identified.
6. At the same time, global connections in the industry have to be intensified, at all levels:
   - education,
   - research,
   - technology and supply,
   - design and market.

 GLOBAL KNOWLEDGE FLOWS AND ECONOMIC DEVELOPMENT – ISBN 92-64-10766-5 – © OECD 2004

7. The entrepreneurial vehicle has to manage the process of sustaining a balanced organisation and financial structure as it, and its value chain network, enter a period of rapid market growth. This growth challenge, or "gusher" experience, is particularly severe for micro-small enterprises and for those in small economies.
8. The region needs to build the cluster and associated infrastructure, so that the externalities can be captured, and a virtuous cycle established. This assists in the winning of international recognition for the enterprises and the cluster, and assists in the voluntary strengthening of the networks in terms of:
   - attraction of students and graduates to the regional cluster,
   - research collaboration networks,
   - global industry development and technology networks,
   - attraction of angel and venture capital investors,
   - supportive regional investments in infrastructure and special facilities and systems.

This is simply a framework, and these are largely outcomes. How they are achieved in practice varies substantially, reflecting differences in industry structure, local resources, local leaders and entrepreneurs. And as the Silicon Valley and Scandinavian exemplars illustrate, there is sometimes the opportunity to take the risk, and create completely new vertical global markets.

Recent New Zealand initiatives in knowledge internationalisation processes

Wood processing strategy

New Zealand is currently in the middle of a several-fold increase in the sustainable harvest of its plantation (mainly pine) forests. This increase in wood volumes is occurring over a relatively short period. The industry did not have a capacity to handle the processing or marketing of this wood, especially of the pruned logs intended for high value markets. There was a high risk of a commodity export trade in logs being adopted.

The plantations are spread over the length of New Zealand, and wood from different regions has been found to have different properties. Export market contacts had been made to explore opportunities for the sale of products processed through expansion of existing plants. These had proved ineffective. Yet when the industry made contact with senior in-market professionals and leaders, it was found that the New Zealand timbers from some regions had highly desirable properties for certain markets, including

Europe. However, New Zealand technologies and processes were designed for other markets, such as Australia and the US.

The inappropriate product specifications and processing technologies employed in the New Zealand industry had led to the export market refusals. However, the market was not saying "No" to the concept of a New Zealand sourced product – just to that being offered. European companies have met with New Zealand industry leaders to design new product and process specifications, and to help New Zealand acquire the appropriate technology.

New Zealand has now established the appropriate global connectedness, and a new niche had been identified for it in the global market place. But as CANZ have reported, a "second generation" of the product was required to meet market requirements. The replication of the CANZ results on a regional and national industry scale had led to proposals for a "rainmaker" programme creating high-level global connectedness for New Zealand industries and clusters. The aim is to secure much higher value-added positions in global industries.

The strategy is being reinforced by a new Beachhead programme of Trade New Zealand. This started with an office for the New Zealand marine export industry in Fort Lauderdale, and the next is being opened for the ICT industry in Palo Alto in Silicon Valley.

New Zealand has also adopted a diaspora strategy, tapping into one million Kiwi expatriates. A private trust, Kiwi Expatriates Abroad, or KEA, operates a referral service for New Zealand companies to high-placed Kiwi executives in overseas markets and industries. Industry New Zealand developed another programme, World Class New Zealanders. This scheme supports New Zealand businesses or a network of businesses to access key international business, market and technology leaders, even if they are not Kiwis. This is a more general access version of the Rainmaker concept, and is heavily oversubscribed.

Research and technology networks

A review of the first 10 years of one of the New Zealand CRIs, Industrial Research Ltd (IRL) (Parker, 2002) noted that it had growing ties in research with universities in joint projects and specialist sub-contracting. Revenues had increased by 50%, and exports had grown from marginal levels to $15 m pa, and were expected to grow to $20m in 2003. International collaborations were growing, and the quality of the science was rising. IRL was engaged in research consortia with other research organisations and the private sector.

IRL had also enjoyed a major increase in patents and license fee income. But most interesting, it had completed some commercialisation activity that resonates with the cases considered above, namely:

- Spinouts of projects with key staff and technologies in areas such as earthquake isolation bearings, superfine chemicals and seaweed and pyromellitate derivatives.
- Contract "current Good Manufacturing Practice" facilities for fine biopharmaceuticals and for chemical synthesis (similar to the strategies used by in Taiwan to form TSMC, and its dynamic semi-conductor industry).
- Transferred technologies to local operations of overseas and to domestic companies.
- Became an integral member of the global high temperature superconductor (HTS) industry as a result of key patents. This includes a strong partnership with American Superconductors Corporation from the US East Coast. HTS is one part of IRL's involvement in the New Zealand nanotechnology project.
- Young scientists from around the world are now requesting to join IRL's teams and become involved in its scientific programmes.

Incorporation of the institute has undoubtedly released a new spirit of freedom and international engagement. Importantly, IRL has been able to build a team to develop global industry and research collaborations to give it recognition and authority in its areas of activity. As a result, it has been able to attract the attention and support of Kiwi scientists who hold leading international research, academic and business management positions. A virtuous cycle of two-way knowledge flows is emerging. Other scientists overseas are taking note and participating.

Education

The provision of education to overseas students has long been a significant industry in many countries, including the US, and Australia. The New Zealand education sector has made significant internationalisation and innovation advances over recent years, including:

- Involvement in research consortia with other universities, CRIs and the private sector.
- The introduction of explicit innovation and commercialisation processes, especially:
  - ❖ Uni Services at the University of Auckland, a proactive innovation search activity across university faculties and institutes, and manager of the commercialisation process.
  - ❖ The Centre for Innovation at the University of Otago, a bespoke wet lab facility for the biotechnology sector, which incorporates:
    - A new innovation search and management activity.
    - An incubator.

  - Corporate lab research suites where major companies can locate research staff to work on joint projects with university staff.
- The emergence of international student education as a major economic market for New Zealand. In 2002, enrolments grew by 50% to 82 000 international students with 41 700 at language schools, 15 300 at secondary schools, and more than 20 000 enrolled in tertiary institutions. Fees and accommodation costs alone amounted to $1.7 bn in 2002.
- The introduction of Performance Based Research Funding for tertiary institutions has commenced, and 25% of the membership of the 11 panels will be from overseas. This is part of a reform of the tertiary sector to build regional and national centres of specialisation related to local industries, and eliminate excessive duplication which has emerged under the previous enrolment-driven system.

Skills and immigration

New Zealand's unemployment rate has fallen below 5%, despite an increase in labour force participation rates. Businesses have been experiencing a skills crisis of increasing intensity in recent years, and permanent net migration figures show a record net inflow of over 40 000 pa. To assist, companies of standing in key industries may accredit themselves to the New Zealand Immigration Service; thereby enabling them to offer positions in the confidence that residency will be granted if agreed conditions are met. This scheme is working well for companies accredited to it.

New Zealand's immigration stance is somewhat more pragmatic than many others (including Australia). It is beginning to display the rich interactions that the Asian countries enjoyed in their relationship with the US West Coast. Some examples are:

- The establishment in New Zealand of a software centre of the University of Washington's Human Interface Technology (HIT) Lab at the University of Canterbury. A research consortium, the New Zealand HIT Lab is built around the return of one of the leading researchers to his home country and Seattle's interest in tapping into a different and distinctive creative culture. Less than a year old, the consortium has already attracted participation from local businesses, and from large MNCs who were active in the consortia at the Seattle lab.
- Scott Technology, a New Zealand based and a leading manufacturer of automated appliance industry production systems has been active in China through joint ventures with its US customers. Last year it completed its first stand-alone contract in China, and decided to open its own office in Shanghai. The manager is Dr Henry Pan, who did an MBA at Otago University, which included a study on the company. He liked it so much he

went to work for it. Over a 4-year period he built up the relations between Scott Technology and the potential client base in China.

- Yihuahi Gao was brought to New Zealand 1992 by Landcare, a CRI. In mid-2002, he left and established his own business, Alpha Healthcare with the support of Crop and Food (another CRI) and Massey University. The company uses New Zealand technology to extract concentrates from Lingzhi mushrooms grown under contract in 10 farms in China. New Zealand offers a technology base, and a highly reputable source for the extracts that go into health supplements. Research is underway to extend the process to products such as cherries, strawberries, kiwifruit, avocado, herbs and marine plants. It is unlikely that this market niche would have been identified without Mr Gao's cross-cultural insights, and market and technology connections.

Investment funds

The increasing levels of commercialisation and innovation activity in New Zealand have attracted the attention of angel investors and venture capitalists from overseas. To build New Zealand capacity in these dimensions, government has established a Venture Investment Fund, which provides a base-line investment which private sector managers then add private funds to, and manage a combined portfolio. With Industry New Zealand support, a business angel investment activity has been established by Industrial Research Limited.

The overseas angel investor networks have been developed under the sponsorship of the network of VECCI – Victorian Employers Chamber of Commerce and Industry. A key ingredient of their service is matching industry experienced investors with new and potential high growth companies. The international market and technology networks that these investors bring give the enterprises connections to the global industry far more quickly than they could ever achieve alone. The angel investor industry itself is highly networked and thus offers a powerful resource for the innovation and internationalization.

Despite these encouraging improvements there are still concerns about the limited capacity of the New Zealand share market. There are strong hopes within the industry that the government's new capital fund for part funding of national superannuation will have substantial local investments, and will breathe new life into the market.

Implications for the Scottish context

Inward investment

Scotland's economic performance is below the UK average. It faces strong peripheral forces as a result of the concentration of population, investment and attention on the South East and the M4 corridor. The 2004 accession of the East European states to the EU shifts Scotland further away from the European market's centre of gravity. The new member states are already attracting new FDI and existing jobs from Scotland.

The legacy of past FDI strategies has given Scotland a very large ICT sector, but one where MNCs dominate the industry and its exports. Unlike the Asian Tigers, the FDI strategy did not commit itself to building domestic market leaders. Many parts of the industry could be characterized as "branch plants", and thus susceptible to the relocation decisions that Scotland has experienced. The Alba Centre has been developed to address this issue.

Without the depth of local company involvement, it seems clear that the essential "global industry connectedness" has largely been executed through the MNCs. Thus, there has been little flow through to local firms. This disconnection, may explain the criticism of the speakers at the conference on "Scotland in the Global Economy", which the review panel participated in, that local management is inward-looking, and not good enough to deliver Scotland's goal.

But Scotland is going through a major transition. Its financial sector has already become an international success story. Centres of acknowledged excellence have been developed in biotechnology (*e.g.* Dundee and Edinburgh) and in ICT at the Alba Centre. The FDI strategy has been radically re-shaped to focus upon high value jobs, within a context of industry building. Disinvestments and job lay-offs in "branch plants" are off-setting progress being made elsewhere. Over time, this negative dynamic should abate, and the underlying progress show through.

Within the new FDI strategy, concern was expressed as to whether Scotland was capturing the available spill-overs. Collaboration with the MNCs with substantial commitments to Scotland has been upgraded, and the "hands off" approach of earlier British policy laid aside. The Lanarkshire Local Enterprise Company (LEC) also demonstrated that corporate spin-outs were a very successful strategy. Enrolment of the MNCs in the three new Intermediate Technology Institutes (ITIs) will also create a significant opportunity for inwards knowledge spill-overs.

 GLOBAL KNOWLEDGE FLOWS AND ECONOMIC DEVELOPMENT – ISBN 92-64-10766-5 – © OECD 2004

Cross-border alliances

Perhaps the biggest problem that Scotland faces is the low engagement of indigenous companies in international markets, and the low rate of commercialisation of the research from its tertiary institutions. There are possibly several dimensions to this issue:

1. Knowledge is not being packaged into "innovations" that can then form the basis for an initial international market foray, or sowing. Scotland is possibly now under-performing here. This may be because the new industries are using completely new technologies, and those who can translate discovery into marketable and deliverable propositions are not yet present in Scotland, or the "connectedness" to the global industry is insufficient. Possibly, technical and design professionals are under-represented. Whatever, the innovation system appears to have a blockage.
2. Scotland has a low ranking as an entrepreneurial business culture. This suggests that there are too few commercial vehicles being created for carrying new intellectual property and innovation. Thus, insufficient Scottish knowledge is receiving international market exposure. Scottish Enterprise and the universities have both undertaken major initiatives to address these issues, including enterprise scholarships and new case management and spin out programmes.
3. The small number of indigenous firms engaged in global markets suggests that Scotland's connectedness to global markets may still have an "exporter" mentality, rather than an "industry shaping" ethos. In part, this reflects the situation in (1) and (2) above. The MNCs with major ICT development and regional support activities based in Scotland will offset the impact of this shortcoming to some extent, but their engagements are largely within their own commercial groups.
4. To the extent that Scotland has a high proportion of micro-small enterprises, it faces a business population that is unable by themselves to enter international markets, or formal research projects. Yet, apart from the ITIs, networking solutions such as clusters and export networks were not presented to the OECD delegation.
5. A strong case was made that profitability of MNCs in Scotland was 60% higher than that of local companies, and for Scottish units of US based MNCs, 90% higher. Again, no reference was made to cluster strategies whereby externalities can be captured to gain such higher returns. These high returns in overseas localities may be one cause of the low penetration of local enterprises into the MNC supply and production systems.

Scottish Enterprise has undertaken a number of initiatives in this regard (see Appendix 1). First, the CONNECT programme has been adapted from San

Diego. Second, a Proof of Concept fund has been introduced to take institutional research projects further towards commercial applications. Finally, ITIs have been established to promote applied research in a consortia-type membership system. The ITIs offer opportunity for pro-active commercial development of innovations by dedicated commercial teams. Recruitment success here may help alleviate the difficulties raised in (1) above.

In the international dimension, there are the STAR sites in the US, which received favourable assessments from the private sector. Angel investor networks are operating. Potential high growth companies are assisted through the Global Companies programme. A global network of expatriate Scots and friends of Scotland has been established with the 500 member Global Scot network.

There are some issues here however:

- Global Companies is reported to be a one-off engagement in highly resourced strategic business planning, with a mutual learning process through the sharing of experiences as after care. In New Zealand, the Fast Forward New Zealand programme is designed to be an on-going case management support aimed particularly at assisting the firms through the "gusher" experience, and lifting the overall survival rate.
- Global Scot is restricted to "family and friends". While it has the strength of being account managed, it does not aim to access the "industry shapers" in the way that the World Class New Zealanders or Rainmaker projects aim to. This may mean that the level of global industry connectedness for Scottish industries will fall below that needed to access the desired value points in international industry.

Attracting talent

The first source of talent is the national population. The Scottish Minister for Enterprise noted a new £300m Skills and Education Strategy had been launched. But other speakers noted that many young Scots were still leaving school early. We are unable to comment further, except to note that New Zealand has felt it necessary to run an Enterprise Culture Programme, and to have a public education programme on innovation.

There are two other major sources for attracting talent to Scotland – local enrolments of overseas students, especially post-graduate, and immigration of skilled people. The essential mechanism for immigration is demand pull from local businesses. This will happen when sufficient Scottish businesses achieve the size, capability growth opportunities which demand more highly skilled personnel to profitably employ a fast-growing number of high skilled personnel.

 GLOBAL KNOWLEDGE FLOWS AND ECONOMIC DEVELOPMENT – ISBN 92-64-10766-5 – © OECD 2004

Many international students are enrolled in Scottish universities. Many do projects that create opportunities for them to stay on in employment in Scotland – but until September 2002 this was a difficult thing to do. Concern remains that UK immigration rules are still too restrictive for Scotland's needs.

Scotland is reported as having relatively few, and rather small, ethnic communities living within it. This situation may therefore limit the scope for the types of industry and market networking and building that the Asian nations and Silicon Valley achieved, and which are also opening up for New Zealand in China and other Asian countries. Given that Asia is expected to be the fastest growing economy this century, Scotland could consider a stronger educational marketing effort in Asia. Further adjustments of visa regulations may assist in this process. Such changes would enable Scottish Enterprise to build a stronger business momentum from its relationship with Singapore.

Other aspects

Intermediate Technology Institutes (ITIs). The three new ITIs represent a major new initiative for furthering the development and internationalisation of Scottish knowledge. They are a network strategy, and provide a vehicle for proactive commercial management of the resulting intellectual property and innovations. The institutes are also designed to build "connectedness" to other global centres of knowledge. Absorptive capacity should be developed in local firms if Scotland is to secure the economic development potential of the ITI investments.

Innovation and venture financing system. In the briefing to the OECD delegation, comments were made that:

- University spin-outs were not developing as fast as corporate spin-outs, with many getting stuck at around 10 employees in size.
- Angel investors were not finding venture capital investors picking up their companies and so had to undertake second-round financing.

These factors suggest that, like New Zealand, the linkages in the Scottish innovation and venture financing system are incomplete. Certainly, the necessity for angel investors to make second-round financing means that the growth path is sharply slowed below its potential. It probably also means that the angels are concentrating in areas other than their specialty, and cannot refresh their portfolio with more spin-outs etc.

If the financing system for innovation and ventures is either weak, disconnected, or in the very early stages of maturity, this may be a significant factor in the slow growth performance in Scotland. Further investigation is needed to understand the situation in this part of Scotland's financial system.

Conclusion

The lack of an "indigenisation" strategy in earlier FDI policies has created something of a present-day penalty for Scotland. But the country has moved quickly to correct these issues. The international experience canvassed in this chapter underlines the critical inter-dependence between innovation and the internationalisation of knowledge flows. Building the local business base within key industries seems to be the key priority, although a number of policies and programmes have to mesh together to achieve this outcome.

Intellectual property/technology packaging is suggested as one area warranting further consideration, and could be a possible target for FDI and/or talent recruitment. Entrepreneurial strategies need to be monitored to ensure that the new programmes deliver. The focus for global connectedness may need re-alignment from the "Scottish family and friends" towards accessing global industry shapers. The ITIs have the potential to be a major enhancement to Scotland's innovation system.

There may be some significant issues in Scotland's innovation and venture financing system. The briefings to the OED delegation pointed to a likely disconnect between angel investors and venture capital and later investment partners. If so, this disconnect could be a major restraint upon the growth of Scotland's new-knowledge enterprises, as it appears to have been in New Zealand.

Notes

1. "Silicon Valley and its Imitators" ...See for example: Bresnahan T, A Gambardella, A Saxenian (2002) "Old Economy" Inputs for "New Economy" Outcomes: Cluster Formation in the New Silicon Valleys". Paper to the DRUID Summer Conference on "Industrial Dynamics of the New and Old Economy" Copenhagen/Elsinore, 6-8 June 2002.
2. Diaspora strategies have become an increasingly common strategy for achieving the "connectedness" goal. In some developing countries, the Diaspora is also tapped as an important source of investment capital. Thus India, with 20 m of its nationals working overseas, hosted a lavish 3 day conference in New Delhi in January for leading members of its Diaspora. There were 2 000 delegates from 63 countries, in the largest event of its kind to date. The focus of the event, plus accompanying Government reforms, was to attract more investment to India, including from its expatriates.

References

Campbell-Hunt, C. (2001), *World Famous in New Zealand: How New Zealand's Leading Firms Became World-Class Competitors*, Auckland University Press: Auckland.

Carlsson, B. (1996), Differing Patterns of Industrial Dynamics: New Zealand, Ohio and Sweden, 1978-1994, *Small Business Economics*, 8, pp. 219-234.

Chapple, K. (2002), Social Innovation and the New Economy: Lessons from Silicon Valley and the Bay Area, OECD LEED Programme, Paris.

Institute of Technology Management (1997), Strategic Technology Management in Korea, Heriot-Watt University: Edinburgh.

Knuckey, S. and Johnston, H. (2002), Firm Foundations 2002, Ministry of Economic Development: Wellington.

Lecuyer, C. (2000), Fairchild Semiconductor and its influence, Ch. 8 in Lee, C.M., Miller, W.F., Hancock, M.G. and Rowen, H.S. (eds) *The Silicon Valley Edge: a Habitat for Innovation and Entrepreneurship*, Stanford University Press: Stanford.

Lee, C.M., Miller, W.F., Hancock, M.G. and Rowen, H.S. (2000), *The Silicon Valley Edge: a Habitat for Innovation and Entrepreneurship*, Stanford University Press: Stanford.

Mathews, J. and Cho D.S. (2000), *Tiger Technology: The Creation of a Semiconductor Industry in East Asia*, Cambridge Asia-Pacific Studies, Cambridge University Press: Cambridge UK.

OECD (2002), *High-Growth SMEs and Employment*, OECD: Paris.

Parker, S. (2002), The Littlest Clue: the resurgence of New Zealand science and technology, Industrial Research Ltd: Lower Hutt.

Rowen, H. (2000), Serendipity or Strategy in C. M. Lee *et al.* (eds) *The Silicon Valley Edge: a Habitat for Innovation and Entrepreneurship*, Stanford University Press: Stanford.

Saxenian, A. (1996), *Regional Advantage: Culture and Competition in Silicon Valley and Route 128*, 2nd Ed., Harvard University Press: Cambridge, Mass.

ISBN 92-64-10766-5
Global Knowledge Flows and Economic Development

Chapter 7

Policy recommendations and learning models

by

Jonathan Potter (LEED Programme, OECD)

This concluding chapter synthesises the main policy recommendations and learning models from the book. It examines in turn the three core study issues of promoting cross-border alliances, stimulating knowledge transfers from foreign investment and attracting talented labour from overseas. The need to strengthen domestic innovation systems as a pre-requisite for connection to global knowledge flows is also stressed.

Introduction

This chapter draws out the main messages from the book, focusing on policy recommendations and programmes providing useful learning opportunities for governments and development agencies. The chapter tackles in turn the issues of promoting cross-border alliances, stimulating knowledge transfers from foreign investment and attracting talented labour from overseas. It concludes with some wider observations on the need to build the foundations for participation in global knowledge flows by strengthening domestic innovation systems.

Promoting cross-border alliances

Policy recommendations

Building a broader base of SMEs involved in cross-border innovation alliances

One of the first challenges for a country or region seeking to promote the participation of SMEs in cross-border innovation alliances (such as international product licensing or alliances for product, process or market development) is to increase the number of SMEs acting at international level. But in order to participate successfully in strategic alliances at international level, SMEs must have strong internal innovation capabilities. These capabilities are required both to generate innovations that can be exploited overseas and to absorb and use innovations generated overseas. If the public sector is to intervene effectively to increase the number of SMEs in cross-border alliances it is therefore important to understand the barriers that SMEs face in innovation. A number of barriers are likely to constrain the innovation potential of SMEs. These include:

- *Lack of funding*. Problems finding risk capital combined with a high risk of innovation in SMEs makes it difficult for SMEs to raise funding for innovation.
- *Limited management resources*. A lack of personnel with higher education and problems recruiting such personnel make it difficult for SMEs to develop and implement innovation strategies.
- *Limited technological competence*. Difficulties in accessing frontier innovation going beyond experience-based, tacit knowledge makes it difficult for SMEs to exploit innovation.

 GLOBAL KNOWLEDGE FLOWS AND ECONOMIC DEVELOPMENT – ISBN 92-64-10766-5 – © OECD 2004

- *Narrow customer focus*. Because the innovation process in SMEs is often dependent on clients, a narrow customer focus and often a lack of feedback from clients makes it difficult for SMEs to access innovation.
- *Lack of time of managers*. The lack of time of managers limits the ability of SMEs to develop forward-looking projects.
- *Poor strategic capability*. Poor strategic capability reduces the ability of SMEs to identify and carry out important innovation projects.

A number of individual policy instruments can be developed to address these barriers:

- Lack of funding may be tackled by financial support for innovation projects and innovative new firms.
- Limited management resources may be tackled by mobility schemes (for example between higher education and industry) to recruit higher educated personnel to SMEs.
- Limited technological competence may be tackled through technology centres providing services to firms.
- Barriers related to narrow customer focus, lack of management time and poor strategic capability can be addressed with policies aimed at making firms aware of the significance of innovation and improving their way of organising work.

However, it is important to design a comprehensive approach to SME innovation support rather than develop individual instruments without regard to their interconnections. In this spirit, two key overall directions for future policy development can be identified. Firstly, an integrated set of measures is required to tackle the innovation barriers identified above. This can be thought of as a strategy to encourage SMEs to "learn to innovate". Secondly, it is important to overcome bottlenecks affecting SMEs in the wider domestic innovation system.

The idea of encouraging firms to "learn to innovate" focuses on increasing understanding within SMEs about the importance of innovation and how to introduce it into their existing routines, rather than simply providing scare resources. A package of measures is therefore required, including instruments to create demand from SMEs to undertake initial innovation activity, to diffuse knowledge about how to perform R&D and innovation and assist firms to make contact with relevant knowledge organisations outside their region. Thus from a "learn to innovate" perspective, financial incentives, for example, must be tied to attempts to change firm behaviour, particularly in SMEs that have not seriously adopted innovation routines in the past. This emphasis on "learning to innovate" is particularly important given that economies are increasingly open and therefore a lot of the necessary technology can be purchased overseas as long as SMEs have sufficient innovation absorption capacity. Innovation capabilities within firms, skill levels and an innovation-supporting institutional framework are critically important to SME innovation absorption capacity.

In order to overcome bottlenecks affecting SMEs in the domestic innovation system it is important to look beyond currently predominant structures for supporting individual firms towards the development of more comprehensive innovation systems stressing interactions amongst firms and organisations. Traditional innovation systems were initially developed around a linear model of innovation, in which large firms, universities and research laboratories were seen as the generators of knowledge through basic R&D, which large firms themselves might exploit or which might filter down to SMEs competent in the commercialisation process. It is now recognised that innovation systems are more complex and that innovation is best stimulated through constant interactions and networking between firms, universities and research laboratories in which even non-R&D-intensive SMEs can participate. The systemic, networked approach to innovation systems brings together governance mechanisms, universities, research institutes, technology transfer and training agencies, consultants and other firms acting together on innovation matters as well as promoting a socially-interactive culture. In particular, there is very often a need to increase the interactions between agents within national and regional innovation systems and to make these interactions accessible to SMEs.

The concept of an innovation system is a broader one than that of a local enterprise cluster. Nevertheless, in selected sectors, cluster building is an important complement to the development of a wider innovation system across the whole economy. Thus it appears to be easier for SMEs to internationalise if they form part of a geographically-concentrated cluster bringing together local suppliers, competitors, customers and enabling conditions such as specialised skills pools and technology institutions. Being part of a cluster assists in winning international recognition for the enterprises in the cluster and in attracting students and graduates, developing research collaboration, industry development and technology networks, attracting angel and venture capital investors and developing supportive investments in infrastructure and special facilities and systems. In order to grow a cluster, it is important to develop new innovations that are complementary to existing technologies in the cluster, stimulate new firm formation and firm growth, develop highly skilled technical labour and managerial leaders, make connections to markets, and encourage cross-border co-operation and competition and capital provision in the given sector. One way of developing connections to

 GLOBAL KNOWLEDGE FLOWS AND ECONOMIC DEVELOPMENT – ISBN 92-64-10766-5 – © OECD 2004

international markets is by helping to position cluster companies into areas not covered by existing world market leaders.

Improving capacity in "learning to innovate" and strengthening national and regional innovation systems and clusters are necessary first steps to building a broader base of SMEs involved in cross-border innovation alliances.

Growing global companies

As well as increasing the number of companies becoming involved for the first time in cross-border alliances, it is important to grow companies that have the potential for gaining some real global reach. A number of recommendations can be put forward on this issue. However, it is interesting first to recall the weaknesses described in chapter 6 concerning the growth of global companies in New Zealand because many of these problems may be common to other countries or regions. The main problems in New Zealand are as follows:

- Growth companies operating in the international market place are taking too long (often 10-15 years) to reach a size and level of engagement where they have recognition and credibility. This compares with targets of "within 5 years" typically used in North American venture and development capital circles.
- The business models adopted by new technology spinouts and new initiatives of firms tend to be managed in an incremental innovation strategy, rather than in the context of radical innovation.
- The business plans submitted by new technology and fast-growing businesses are often limited in their appreciation of and strategies around international competitors and the business models and value-chain structures of the international markets they are engaged in.
- The venture and development capital strategies for investors tend to imitate those of North America (aiming at an Initial Public Offering, or IPO) when the reality is that most of New Zealand's exits are made through "trade sales" to overseas companies.
- New Zealand's enterprises are predominately micro or mini in scale and their domestic market is small. Thus, their challenge in scaling up to internationally relevant and robust stature and capability is greater than for counterparts in larger economies.
- The domestic capital market is small and has not attracted sufficient international investor involvement to provide a viable basis for IPO investor exit strategies.

The above issues may help explain why New Zealand fast-growing enterprises engaged in international market-building have so frequently been sold to overseas counterparts. Two particularly important factors appear to be the extra degrees of scale that a firm must build up from the domestic base to become an international market leader and the severe limits on domestically-financed expansion strategies.

A small firm seeking to enter global markets from a comparatively small domestic market is likely to experience rapid expansion and overseas market growth, which will be difficult to manage. In the case of New Zealand, the transitions that globalising firms undergo can be summarised as:

- Development of a successful innovation.
- Initial engagement with international markets in a "sow and reap" strategy which inexpensively gains an appreciation of market potential and the need for design and other modifications to meet market needs and competition.
- Lifting (or often continuation) of high level R&D activity, typically at around 10% of sales, and enterprise focus on a promising product or market.
- A period of very rapid growth (the "gusher experience"), typically flowing from a second generation of the product or service.
- Keeping the enterprise "in balance" as its marketing, production, logistics, training and management systems are rapidly expanded.
- Completion of the transition to a "future capable company".

Often the innovation comes from a small enterprise operating in low-margin market segments at the bottom of the value curve. The transition to higher value/higher margin points on the value curve is essential to provide the equity returns necessary for growth, but this transition is not automatic.

SMEs in other countries and regions are likely to face similar barriers to international growth and a similar, if perhaps slightly less severe, transition experience. Policy support should therefore focus on trying to assist companies that are capable of growing globally to overcome the barriers they are likely to face and to successfully manage critical transitions in their growth processes. The experience of New Zealand suggests that companies growing into global markets benefit from an overseas partner or advisor. Such partners can help the SME with developing and tightening its product range, market positioning, market proposition, technology acquisition and development and material and components supply, complementing advice received from overseas customers. Policy should therefore aim to help growing companies to identify suitable overseas market or technology partners in their industry. Emigrant networks overseas can be very helpful for firms wishing to establish overseas strategic alliances. In addition, companies growing into global markets need to be innovative organisationally, undertaking internal process innovation, business reorganisation and marketing innovation. Therefore policy to encourage exporting by fast growth SMEs should be accompanied by measures to make firms aware of the organisational implications of growing into overseas markets and to encourage necessary organisational change, especially during the "gusher experience".

Learning models from Scotland

The Proof of Concept Fund

The Proof of Concept Fund aims to fund early stage research work that will help develop a technological innovation towards eventual commercialisation. It was launched in 1999 and now has a budget of £33 million available over a six year period. It helps researchers in Scotland's universities, research institutes and National Health Service Trusts to take pre-commercialisation projects further towards commercial applications.

The fund currently supports 120 projects. These projects occur after advances have been made in strategic research or a background patent has been filed. These ideas should have the potential to form the basis of either a new high growth business or a license to an existing company. However, it is recognised that the supported projects are high risk and may take several years to become commercially attractive to private investors. The Proof of Concept Fund intervenes to support the further development of innovative ideas in the period before researchers are able to produce the full lab-scale demonstrations of their technology or pre-production development and prototyping that could attract private investors.

The role of the Fund is therefore to help grow Scotland's future knowledge-based companies, those that will be able to exploit global technology niches and contribute to the knowledge-out part of the Global Connections Strategy.

Global Scot

Global Scot was established in 2001 as "a global network of influential Scots and those with an affinity for Scotland who can contribute to and share in Scotland's economic success". It has approximately 650 members (August 2003) of which some 39% are from Europe, including a small number in Scotland, 49% are from North America and 12% from the rest of the world. The members are in key sectors that are important for the future development of Scotland (*e.g.* electronics and communications technology; finance, accountancy, law and business services; biotechnology and life sciences; and creative, media and consumer industries). Key individuals are selected for First Minister level invitations to join the network. These individuals must work in a key sector for the Scottish economy and hold a senior or influential position.

Key activities of the network are events, a website and "member roles":

- *Events*: Events are held on specific issues (*e.g.* developing the Scottish venture capital market, commercialisation of Scottish research, marketing Scotland overseas) held for example in New York or Silicon Valley.
- *Website*: A web-based tool enables direct communications between members and can be used as a tool for their personal networking. The website includes

news headlines from Scotland, information on industry trends in Scotland, information about member events, member success stories and a directory of members. Members can contact each other using this web tool; they can also select their level of "visibility" to enable anonymity from other members if preferred.

- "*Member roles*": Members can volunteer to take on specific roles, such as business support, intelligence support, school support and economic development support. These "member roles" are agreed after consultation and piloting with Scottish Enterprise. Examples of such hands-on activities are providing world class advice and sector and market intelligence to Scottish Enterprise, mentoring Scottish companies in developing alliances overseas, speaking on public platforms about Scotland, such as at trade missions at major trade fairs, adoption of Scottish schools for mentoring and entrepreneurship education, advice on raising venture capital for Scottish companies, and advice on branding Scotland for future business promotion.

Key factors which help explain the success of the initiative to date include the following:

- The network is sold as belonging to the members not to Scottish Enterprise.
- Activities are business focused and address the professional interests of the members.
- The network provides people who wish to contribute to Scottish economic development with a structured means to do so.
- The network provides easy access to global peers.
- It is a source of knowledge, opportunities and ideas for the members.
- There is a range of tailored information services for the members.
- There are issue focused events.
- Each member of the network is assigned a senior support manager from Scottish Enterprise to provide service support.

Significant work is required to identify and invite key contacts, consult with the contacts (often carried out in person by Scottish Enterprise field officers overseas), mobilise the contact and match them with activities where they could best assist and to undertake sustained engagement and tracking of the members' activities.

One of the many benefits of the Global Scot network is in assisting Scottish firms to enter into cross-border strategic alliances, either through advice and mentoring, or provision of contacts, or assistance with finance. In addition, the network provides intelligence to help Scottish Enterprise develop its strategies and programmes and improve the image of Scotland overseas. It is a relatively low cost programme.

Global Companies Development Programme

The Global Companies Development Programme targets companies that could become significant players in the global economy and both want and could benefit from Scottish Enterprise support. Approximately 25 companies were taking part in 2002, all of which were committed to accelerating their process of internationalisation over and above exporting. The main purpose of the programme is to develop the strength of the management team to meet the challenges such international expansion will impose. To do this participating companies receive bespoke consultancy support for a period of 3-6 months as well as access to continuous and ongoing peer group learning and specialist events.

Scottish Enterprise developed responses to three major economic concerns to Scotland in the 1990s: the low business birth rate, weak commercialisation from the indigenous science base and the need to respond to the challenges posed by globalisation. The Global Companies Development Programme emerged as a direct response to the third of these issues, acknowledging that Scotland has fewer global companies than it would like and proportionally fewer than benchmark countries such as Switzerland and Sweden.

Research by Scottish Enterprise indicated that key barriers to the internationalisation of Scottish organisations included:

- Lack of an international culture, vision and mindset.
- Lack of a coherent internationalisation strategy.
- Shortage of finance.
- Isolation from business networks.
- Lack of support infrastructure.
- Lack of experience in internationalisation.

The Global Companies Development Programme aims to address these barriers through consultancy support, peer group learning and specialist events leading to the development of top class and internationally focused management teams.

Consultancy support. Within the context of developing the management of each participating company in their particular situation (sector, markets, etc.) the consultancy provided by the programme has three key goals: generating a shared vision of the future; developing strategic goals based on the new vision; and developing key actions to deliver the strategic goals. There are five major steps to the process, namely obtaining management team and board commitment, interviews and skill assessments with individuals in the company leading to the development of a vision for the company's development and management requirements, scenario planning on internationalisation factors that could impact on the company, generating strategic options and consequences and preparation of an action plan for the internationalisation strategy. This consultancy allows participating companies to examine and develop the core competencies necessary to accelerate their internationalisation over and above exporting. All the different areas of the business with a potential impact on internationalisation are examined, including strategy and planning, existing international operations, marketing, operations and technology, research and development, organisation and human resource management and finance. The aim is to develop an agreed vision for the future of the company and a strategy for implementing the vision, addressing issues like which markets to address and products and services to develop, how to go about it and when.

Peer learning. The peer group learning sessions involve invitations to the Chief Executives of client companies to quarterly meetings based around an identified theme and speaker (such as branding or finance). This provides an opportunity for the Chief Executives to form networks and importantly to share experiences of their international development.

Specialist events. Specialist events are also held, to which potential globalising firms are invited, for example on scenario planning or international management skills development.

The Global Companies Development Programme also links with other initiatives. For example, it is involved in the stimulation of bilateral relations with target foreign economic development agencies, including in the Asia-Pacific and North American regions, in order to help develop partnerships between firms in the areas of the respective economic development agencies. It is involved in working more with the academic and financial communities to support commercialisation and finance provision. And it also links with the Virtual and Incubator Offices (see below) and Scottish Development International field offices overseas in order to assist client firms to internationalise.

Overall the Global Companies Development Programme encourages firms to consider and develop approaches to allow a more aggressive expansion into global markets. Further support for specific issues needed to implement the action plan agreed through the programme may be available from the Local Enterprise Companies in the Scottish Enterprise network, which are an integral part of the strategy and delivery process.

Virtual and Incubator Offices (formally known as STAR Centres)

A network of four virtual and incubator offices are currently operating in the USA. One of their key functions is to provide incubation space for Scottish companies to assist them to enter the North American market, with support

 GLOBAL KNOWLEDGE FLOWS AND ECONOMIC DEVELOPMENT – ISBN 92-64-10766-5 – © OECD 2004

of staff from the Scottish Development International field offices. They also allow Scottish companies to operate initially from the centre in a virtual way, by providing a mailbox address in the centre, which helps firms with early marketing. The virtual office concept, with staff answering calls and mail to the international address on behalf of the company, has been extended around the globe.

Learning models from other countries

Technology brokers: the Norwegian TEFT programme

The Norwegian TEFT programme (technology diffusion from research institutes to SMEs) aims to increase collaboration between SMEs and major Norwegian national research institutes. Ten county-based technology brokers visit SMEs to analyse their needs and identify technological opportunities that could be exploited through joint SME-R&D institute technology projects. They then broker the development of networks and joint R&D projects. TEFT brokers work mainly with non-R&D-intensive SMEs, which are seen as the kind of firms most in need of external R&D competence to increase their innovativeness. The brokers also aim to change the attitudes and behaviour of large R&D-institutes towards working with SMEs. The approach is very useful in building the innovation capacities of SMEs and getting them used to working on technology projects with partners.

However, a problem that has emerged in Norway is that because the technology brokers tend to be former research employees in the large research institutes served by the programme they tend to be more oriented towards research institute needs than those of SMEs. In addition, non-R&D-intensive SMEs do not always need high-level research contact with leading research establishments but often may be better served by contacts with specialised consultants.

The following lessons should therefore be kept in mind if an attempt is made to adapt the Norwegian TEFT broker initiative to other countries and regions:

- Ensure user orientation. The principal focus of technology brokers should be on competence building within SMEs rather than project generation for universities or research institutes.
- Professionalise and upskill the jobs of the brokers. There is a need to focus on the ability of brokers to work with SMEs because the success of such an initiative depends very much on the brokers' skills.
- Maintain the focus of the initiative on innovation needs and barriers in SMEs. There should be a conscious effort to avoid the temptation to downgrade to basic business development support.

- Rather than limiting the initiative to encouraging national networking, as is the case with the Norwegian TEFT brokers, the approach could be extended to encourage cross-border alliances by using brokers to identify and encourage global as well as national links.

Mobility schemes: First Enterprise, Wallonia, Belgium

Mobility schemes assist SMEs to recruit highly educated people and to increase their contacts with universities and knowledge organisations. This should result in their innovation activity and capability being strengthened. First Enterprise ("Formation et Impulsion à la Recherche Scientifique") in Wallonia covers up to 80% of the salary for two years of a young researcher working part-time in an SME and part-time in a research laboratory. The researcher spends as much time in the hosting research institute as in the SME in order to allow a real knowledge transfer to the SMEs. The research team as a whole must be competent in the relevant research field and motivated to engage in innovation projects in the firm. This type of approach aims to assist SMEs to "learn to innovate" and may lead participant firms to undertake more research and sustain longer-term links with research organisations.

Nonetheless, the following lessons should be kept in mind in applying this type of mobility initiative elsewhere:

- Programme design and management should counter tendencies for research institutes to see a mobility programme as a method of subsidising researcher salaries on existing work rather than as an opportunity for new work with SMEs.
- Universities, research laboratories and researchers should not be allowed to concentrate on basic technology development at the expense of strengthening the long-term innovation behaviour of SMEs and supporting the commercialisation of innovations.
- Awareness and marketing efforts should be undertaken to ensure sufficient take-up of the scheme by both SMEs and research institutes.
- Programmes could be extended to stimulate links with overseas knowledge sources where appropriate, rather than focusing exclusively on national knowledge sources.

Community networking: the "TechAction" Programme, Ottawa and St. John's, Canada

The City of Ottawa and its industrial association, the Canadian Advanced Technological Alliance (CATA) has put in place the "TechAction" programme in order to tackle cultural issues around risk aversion and inward-looking mentality amongst SMEs. The programme aims to involve local firms and institutions in the preparation of a strategy for the development of the high technology sector in the city. CATA undertook in-depth telephone interviews with approximately 100-150 local Chief Executive Officers, managers of firms and laboratories and university administrators and researchers. It also published one page "write in" invitations in local newspapers for anyone in the high tech community with a view on the strengths, weaknesses and future of the community. Facilitated Town Hall meetings were then carried out, led by the Mayor, with discussions – by the community, for the community – dealing with four key elements: leadership, people, money and infrastructure. These discussions led the high technology community to share views as to who should do what in their collective quest to ensure a vibrant community future and to construct advantage through alliances and shared strategies. As well as increasing local networking, industry associations like CATA are helping to introduce Ottawa firms to counterparts in Tucson Arizona, Austin Texas and Monterrey Mexico.

A similar programme was undertaken in the much more remote coastal community of St. John's in Newfoundland, hundreds of miles from North America's main cities. In St. John's, the key feature of the programme was the "TechAction" Town Hall meeting with the Mayor. In this meeting the audience developed collectively a view of the future of their community in the knowledge-based global economy. The role of Memorial University was seen as key, as were the federal laboratories of the National Research Council. It was agreed that realising their collective futures in petroleum, cold-ocean drilling, marine biology, fish stock management as well as information technologies and geographical positioning systems would require community alliances, both within Newfoundland and across Canada and the world. This initiative therefore promoted a vision of local and global strategic alliances for future technology development in key sectors of the local economy. Leadership from all sectors was seen as key for the success of such alliances.

Export networks in New Zealand

Trade New Zealand (the government trade department) runs a heavily over-subscribed export network programme. The rationale is that few micro or small enterprises can invest the necessary time and other resources to make a sustained commitment to international market development. Often their production capacity is too small to deliver the volumes required for market success. However, these requirements can be achieved through the creation of export networks with other micro-small enterprises, thereby giving them market reach, capacity to meet export orders and the (shared) resources to undertake market research and development. These export networks are often linked to Industry New Zealand (the government industry department) programmes for the development of industry clusters. Cluster building is expected to generate opportunities for the development of viable specialist firms and activities that would not be possible if the micro and small enterprises remained self-sufficient and to enable business networks from within the cluster to enter international markets professionally.

Business angel networks in New Zealand

Problems have been identified in New Zealand in increasing the number of SMEs involved in cross-border alliances, related to a relatively weak venture financing system and poor business planning by SMEs. In particular, there seems to be insufficient planning of the initial global market engagements of New Zealand SMEs and critical alliances are not being made. However, angel investor networks have become increasingly active in helping to overcome these weaknesses over the last five years. Their interest, as well as interest by overseas venture capital providers, reflects increasing levels of commercialisation and innovation activity in New Zealand. To help develop capacity, Industry New Zealand is supporting overseas business angel investor networks under the sponsorship of the VECCI network (Victorian Employers Chamber of Commerce and Industry). A key component of the service involves matching industry experienced investors (normally in New Zealand) with new and potential high growth New Zealand companies. The international market and technology networks that these angel investors bring provide connections to global industry far more quickly than the SMEs could achieve alone.

Tapping into New Zealand's diaspora

New Zealand has approximately one million expatriates living abroad, and is tapping into this diaspora in order to help grow global companies. There is an interesting private initiative run by a trust called Kiwi Expatriates Abroad (KEA). This operates a referral service for New Zealand companies to high-placed Kiwi executives in overseas markets and industries. The government has also developed its own programme through Industry New Zealand, called World Class New Zealanders. This scheme supports New Zealand businesses or networks of businesses to access key international business, market and technology leaders. The programme is heavily oversubscribed. A key learning point for other countries and regions is that World Class New Zealanders do not need to be from New Zealand, allowing New Zealand to identify world experts regardless of nationality or affinity.

Silicon Valley, USA

The development of the Silicon Valley economy since the Second World War is very instructive for the knowledge flow and industry internationalisation issues being considered here. Critical factors in the internationalisation of Silicon Valley SMEs have been research access, commercialisation, venture capital, industry connectedness, growth management, new business models, industry R&D, an entrepreneurial education sector, a networked business structure, a regional technology loyalty and entrepreneurship. The business dynamic in Silicon Valley is one of horizontal, highly competitive industry structures, spin-out practices and strong commitment to entrepreneurship and global business models. In the case of Silicon Valley this highly entrepreneurial climate has largely been created by the market, although government research expenditures were an early stimulus. Other countries and regions may have to be more interventionist to try to stimulate some of the factors that have been important to the success of Silicon Valley.

Finland: External links for firms in regional clusters

There is a body of research that shows that firms in a regional cluster need national and international linkages as well as intra-regional linkages in order to remain innovative and competitive. The Finnish view on this is that fierce global competition forces companies to search for the most advanced university institutes worldwide. To be amongst the most successful companies, firms therefore need to engage in global innovation networks and make use of foreign basic research, exploiting world-class, national and international competence centres and innovation systems. It is therefore argued that national and international networking must be intensified in collaborations between public and private players in order both to utilise knowledge available elsewhere and to improve the innovation services needed by small and medium sized enterprises.

Stimulating knowledge transfer from FDI

Policy recommendations

Building links between FDI, indigenous firms and universities

Many countries and regions lack an explicit programme to improve links between foreign investors and domestic firms and universities. However, such collaborations should be encouraged because they can be an important channel for integrating local suppliers, customers and universities with global knowledge sources and outlets. Encouraging such collaboration is likely to be most successful and have most impact within the strongest clusters of firms in leading-edge sectors.

The extent to which an FDI affiliate will impact on knowledge flows is to a large extent determined by the wider business systems of the parent company. Policy makers therefore need to start by understanding both the foreign subsidiary role within its parent company and the business system of the parent company in terms of what knowledge is flowing where. They need to identify those subsidiaries that act as significant nodes in knowledge flows within the parent firm and that could be important learning interfaces both for local firms and universities.

Nonetheless, a note of caution should be added about pursuing programmes that focus on stimulating the transfer of knowledge from FDI affiliates to local suppliers by encouraging local purchasing linkages. In a globalising knowledge economy, research-intensive firms must interact with specialists of many kinds, and these specialists are unlikely to be located locally. This limits the prospects for building comprehensive regional supply chains. The point can be illustrated with reference to the life sciences industry. In this sector, supply chain promotion might not bring major benefits to economies hosting FDI affiliates because life sciences companies have to source from very specialist and leading research institutes and suppliers worldwide, which are likely to be located outside of the host economy.

In addition, it has often been argued that much foreign investment is often of a "branch plant" nature when subsidies have been used to attract foreign investors to high unemployment regions. Branch plant type operations tend to undertake limited R&D and innovation compared with the rest of the parent company and have limited linkages with local universities and firms, thus reducing the scope for knowledge transfer to the domestic economy. In these circumstances, policy should move to support networks and clusters of firms in which technological linkages and interactive learning are more important. This implies integrating FDI policy with cluster policy by attracting FDI that will fit with existing clusters and by tying it in with local firms and universities in that cluster, rather than pursuing broader subsidy-based promotion with little focus on an area's sectoral strengths. Enrolment of foreign affiliates in intermediary technology transfer agencies would also create a significant opportunity for inward knowledge spillovers.

Promoting corporate spin-outs

There is often scope for increasing knowledge transfer from FDI by encouraging the establishment of corporate spin-out firms. Often governments and economic development agencies feel that they need to be more diplomatic than is actually necessary towards foreign firms, reflected in a certain hesitation to talk to FDI about corporate spin-offs. However, the operation of various FDI aftercare policies has shown that it is possible for policy organisations to discuss a wide range of matters relating to company strategy, including attitudes of firms towards corporate spin-outs. In such discussions, it often turns out that foreign affiliates are not opposed to the idea of a manager or a small team of their managers starting a spinout company. Literature on corporate spin-outs suggests a number of reasons why large companies may gain from spin-outs, including the possibility to buy-in new products that would not provide a sufficient return on investment for

 GLOBAL KNOWLEDGE FLOWS AND ECONOMIC DEVELOPMENT – ISBN 92-64-10766-5 – © OECD 2004

them to develop in-house. If governments and economic development agencies adopted a more proactive position or a programme for corporate spin-outs, then more spinout activity might be more forthcoming.

Learning models from Scotland

The Alba Centre

The legacy of past FDI strategies has given Scotland a very large ICT sector, but one where foreign affiliates dominate the industry and its exports. Many parts of the industry could be characterised as "branch plants", and thus susceptible to the sorts of delocalisation decisions that Scotland has recently suffered from during the global ICT downturn. A centre of acknowledged excellence has been developed in ICT at the Alba Centre, which should help to address these problems by better embedding foreign ICT affiliates in a strong supporting environment and by helping to build domestic competencies that will dovetail with those of foreign enterprises in the ICT sector.

The Alba Centre was set up in 2001, with the vision of developing Scotland as a leading world centre of microelectronic product design and technology. Its core activities are:

- The Alba Campus, in Livingstone, near Edinburgh, which is a 96 acre technological park where ICT sector foreign investors and newly started microelectronics design companies can co-locate. A flagship foreign investment has been obtained from the company Cadence Design Systems. The Campus also includes a microelectronics test centre, which provides a test facility for "system on a chip" designs for Scottish and overseas companies.
- The Scottish Embedded Software Centre (SESC), which aims to act as a hub for embedded software expertise. SESC provides access to embedded software expertise, links to local industry and academic institutions and technical and strategic resources. SESC also offers a brokerage service for enquiries of a non-technical nature, creating links and networks within the industry or facilitating partnerships between organisations.
- The Virtual Component Exchange (VCX), which is a trading platform for intellectual property in semiconductor design. Virtual components are pre-designed blocks of code that can be plugged into a complete "system on a chip" design. Using pre-designed materials helps companies to develop more complex electronic systems and speeds up the design process. The VCX helps developers and users trade these virtual components and deal with the business and legal issues involved. The VCX has created a web-based system to unite developers and users. This offers a structured list of virtual components, tools to create business models and contracts, help to

choose virtual components and a framework to complete deals and contracts. Members of VCX include both foreign and Scottish companies.

- The Institute for System Level Integration (ISLI), which provides a platform and critical mass for postgraduate education, professional training, research and support for the electronics design community. The Institute was formed as a partnership between four of Scotland's leading universities in this field (Edinburgh, Heriot-Watt, Strathclyde and Glasgow). It provides Masters and Doctoral programmes and professional training in fields such as electronic engineering, computer science and semiconductor testing. The Institute also supports the development of new technologies through its research work.

The Alba Centre should help to increase the knowledge-intensiveness of FDI attracted in ICT, improve the retention of ICT based FDI and improve links between foreign investors and domestic firms and research organisations.

FDI aftercare and "Team Scotland"

Scottish Development International (SDI) is a leading investment promotion and trade body (incorporating the former Locate in Scotland and Scottish Trade International) operated by Scottish Enterprise. It places significant emphasis on aftercare and embedding of foreign investment as well as attraction of new projects. Aftercare services seek to secure expansion investment and higher proportions of high value activity (including research, design and development), to reduce the risks of contraction or closure, to tap into new FDI opportunities from existing investors, and to obtain linkages to other players.

SDI's aftercare strategy is to provide a wide range of support services to foreign investors already in Scotland through a formal account management structure drawing on Scottish Enterprise's Local Enterprise Companies (LECs) and numerous public and private sector partners. This idea of partnership for aftercare activity is encapsulated in the label "Team Scotland", which is used to describe the close working between the relevant players. Thus rather than attempting to deliver all aftercare itself, SDI will often provide links to those organisations that are best placed to help the foreign affiliate to overcome a specific problem or to make a specific investment. At the same time, considerable effort is taken to ensure that the approach to aftercare is efficient, that accountability is recognised and understood, that there are shared information systems between the main players, and that bureaucracy is minimised to provide a clear and simple process that clients understand.

There are three principal elements to SDI's aftercare support: *i)* day-to-day local relationships, *ii)* conversion of new projects, *iii)* head office relationships. The aim is to develop a shared vision with the foreign investor by understanding the corporate strategy of the headquarters, the performance of the Scottish facility, and the future development of the Scottish facility. Each company has an account leader who will ensure that a formal account development plan is prepared and updated. Liaison with the parent company will normally be undertaken by SDI field officers in the appropriate geography, and liaison with the Scottish plant will often be undertaken by the Local Enterprise Company in the area of the plant. Foreign investors are classified according to an agreed segmentation process. SDI is more proactive with "Gold" accounts and more reactive with other accounts.

The issue of embedding foreign investors has also moved up the agenda in Scotland in recent years. Although Scotland was very successful in attracting and retaining inward investment in the 1980s and 1990s, it is recognised that there has not been as much positive spin-off for the Scottish economy as initially desired. This is reflected by issues such as:

- Low workforce turnover, so that even though foreign investors were very active in workforce training there is likely to be a limited spread of best practice to domestic firms through labour mobility.
- Limited supplier development by foreign investors of Scottish firms, with few Scottish suppliers becoming suppliers to the parent firm at an international level.
- Limited R&D diffusion, as characterised by relatively few spin-outs or links with universities.
- Very limited corporate venturing.

Although there is no specific, independent embedding programme, the focus of FDI policy has therefore shifted in three ways in order to help better embed foreign investment projects:

- *Targeting high value FDI.* The focus of FDI attraction has shifted towards higher value FDI, which is likely to have greater spin-off potential and be less tempted to relocate to low cost competitor areas. Key targets are research, design and development activities, knowledge intensive services (*e.g.* financial and business services) and FDI to support the growth of the indigenous company base (*e.g.* attraction of venture capital).
- *Propositions based on Scotland's assets.* Foreign investment is also likely to be better embedded in sectors where Scotland has key strengths in terms of local firm networks, specialist skills and a strong research base. There is therefore an increased emphasis on attracting firms in such sectors as well as on attraction of investments that will fill gaps in the supply chain of key sectors. The Alba Centre is an example of an initiative to attract companies into a strong base of research and skills.

- *Working with existing investors.* Scottish Enterprise is also working more with existing investors to encourage them to move up the value chain, to develop their local supply chain and increase their general networking with the local economy and society. There is also a strategy to work with companies undergoing rationalisation in order to retain value in Scotland through consolidation of activities in Scotland, support for corporate spin-outs or, through positive relations, working to capture future investment.

Learning models from other countries

Syntens: The "Knowledge-intensive Industrial Clustering" (KIC) programme, Limburg, Netherlands

The "Knowledge-intensive Industrial Clustering" (KIC) programme in Limburg, the Netherlands, is a joint public-private initiative between the regional development organisation Syntens and the multinational copy machine producer OcE. Syntens and OcE started a pilot project in 1994 in order to upgrade the knowledge base of regional SME suppliers. Syntens considered the lack of innovative medium-sized firms and lack of communication and co-operation between certain large companies and local SMEs to be a bottleneck for further industrial development in Limburg. The programme therefore aimed to improve inter-firm co-operation amongst SMEs working together for a client or lead enterprise. The first project formed seven "micro-clusters", each including 4-5 SMEs co-operating in an innovation project with OcE. The firms started with a drawing or laboratory model of new parts or modules for OcE's new colour copier that could be produced at marketable prices. The programme, and mainly OcE, subsidised the development costs in the SMEs. OcE found the project to be very successful and took over its organisation from Syntens in 1996. OcE's motivation was to upgrade their supplies and to find new and more knowledgeable suppliers to contribute to the company's product development.

This programme demonstrates how an economic development agency can work with an FDI affiliate to encourage knowledge transfer work with networks of SMEs. However, in adapting such a programme to other places it is nonetheless important to provide mechanisms to avoid "lock-ins". These may occur if a programme ends up supporting strong ties between a small core of client firms and their existing suppliers, with few openings for other competent clients or suppliers to join the network. Programme design and management should therefore seek to ensure that there are a number of lead companies and openings for both client firms and SMEs to enter or leave the programme. Universities and research laboratories could also usefully be included in the networks.

 GLOBAL KNOWLEDGE FLOWS AND ECONOMIC DEVELOPMENT – ISBN 92-64-10766-5 – © OECD 2004

Taiwan: ITRI and linkages with multinationals

Taiwan has for some years encouraged knowledge transfers from foreign affiliates in order to build a domestic world class manufacturing capacity in semiconductors and information and communications technologies. One of the critical instruments behind this success is the Industrial Technology Research Institute (ITRI). This is an example of the sort of intermediary technology transfer agency that this book calls for. In Taiwan, ITRI formed research alliances with incoming foreign investors and then commercialised outputs into domestic SMEs. The Hsinchu science-based technology park attracted foreign investors and was a suitable place for alliances to be built. For example, key alliances were developed by ITRI with IBM and Motorola. Thus Taiwan's current dominance in mobile Personal Computers (PCs) rests on the work of public-private consortia that rushed product to world markets in 1991. Its strong performance in data switches, crucial to PC networks, also arose from ITRI-SME consortia to match the Ethernet standard in 1993. The IBM PowerPC microprocessor of 1995 was cloned simultaneously through ITRI's alliance with originating designers IBM and Motorola. Subsequent successes arose via the partnership mechanism in the digital communication and multimedia areas. Today, Taiwan is the world's largest producer of notebook computers and a range of PC components such as motherboards, monitors, scanners and power supplies in keyboards.

In most OECD countries the prospects for commercialising original university research are higher than they were in Taiwan, and it could therefore be argued that there is less need for building alliances with foreign subsidiaries. However, research, design and development undertaken in multinational subsidiaries remains a potentially important knowledge source, together with the basic science and research undertaken in universities. Thus similar intermediary technology transfer agencies would appear appropriate in many other countries in order to provide a number of the practices and services carried out in Taiwan by ITRI, in particular the approach of forming research alliances around key sectors. One of the lessons of ITRI is that such intermediary technology transfer agencies need not focus solely on university commercialisation, but should exploit all potential knowledge sources, including FDI, which has proved so powerful in Taiwan.

The Singapore semiconductor industry

Singapore is another country that has created a successful semiconductor industry on the back of US, European and Japanese FDI investment, and later from FDI from other East Asian countries. Singapore has concentrated on working with these foreign investors to raise its local skill and technological levels. A sign of its success is that Singapore now has a cluster of SME service firms and indigenous wafer fabricators including Chartered Semiconductor, a state-owned company. Singapore encouraged closer FDI relations with local suppliers through its "Local Industry Upgrading Programme". In addition, it liberalised its immigration and work permit systems for foreign professionals (engineers) being employed by foreign firms. Singapore is currently mounting a major internationalisation drive by establishing jointly-owned technology parks in a number of countries, including China, India, Indonesia and Vietnam.

The example of Singapore's semiconductor industry underlines some potential lessons for other countries and regions. First, it shows the value of an explicit programme for encouraging relations between FDI and local industry. Second, it shows how the attraction and development of high value added FDI can be facilitated by easing recruitment of highly qualified professionals. Third, it suggests the potential for more pro-active exploitation of knowledge in foreign companies by the establishment of co-owned science parks in other countries.

Vote Research Science and Technology, New Zealand

Networking strategies have recently been introduced in New Zealand via the Vote Research Science and Technology (Vote RS&T) system, which supports research consortia. The research consortia typically include a number of tertiary education and research institutes collaborating with a group of private sector companies. These companies include a number of local operations of overseas multinationals as well as indigenous large and medium scale enterprises and some specialist micro-small scale enterprises. The involvement of overseas multinationals is seen as important to the research networks in two dimensions. Firstly, they provide access to global research, technology and market networks. Secondly, they offer extensive international market and customer support networks for commercialisation of intellectual property from the joint ventures pursued within the consortia. This government initiative illustrates how FDI affiliates can help domestic SMEs to access global knowledge flows through binding them together in common innovation projects.

Attracting talented labour from overseas

Policy recommendations

Critical mass and economic opportunity

The principal mechanism for attracting talented labour, and hence tapping into new sources of knowledge, is the demand for highly-skilled labour from domestic businesses. This requires a critical mass of domestic businesses demanding highly-skilled personnel. At bottom line, people will be

 GLOBAL KNOWLEDGE FLOWS AND ECONOMIC DEVELOPMENT – ISBN 92-64-10766-5 – © OECD 2004

persuaded to migrate if there are better job opportunities and wages in the host country than in their home country or alternative locations considered. And often highly-skilled people are not looking simply for an area that will provide a first job, but one with a sufficient critical mass of firms in their industry to allow for career progression and learning. Thus, the existence of a substantial number of well-paid job opportunities in given sectors will be a strong motivation for people to migrate. In addition, a significant number of highly-skilled migrants move between sites within multinational corporations, so the attraction and retention of knowledge-based FDI also has an important role in attracting highly-skilled people.

Policy can best influence these critical mass and economic opportunity issues by building up strong innovative and highly productive sectors and clusters through supporting local skills, knowledge institutions, specialist finance, knowledge networks and so on. Indeed, strong sectors and clusters will be important not just for attracting talent from overseas but also for retaining domestically-trained talent. However, as well as building opportunities in critical sectors it is also important to market and publicise them, for example amongst overseas professionals and diaspora networks.

Addressing "people climate"

Recent thinking, led by Richard Florida of Carnegie Mellon University, stresses the importance of a good "people climate" to attract talented people, complementing traditional policies that act on "business climate" to attract businesses. It suggests that in order to attract skilled researchers, workers and managers in high technology and creative sectors, policies for business development should be supplemented by policies for attracting talented people by improving the perceived and actual quality of place. Professor Florida focuses on the need for a place to reflect the lifestyle interests of creative people and for places to be tolerant, diverse and open to creativity. Beyond this, we can look at critical requirements for key groups of people that an area is likely to wish to attract. For academics and university researchers, the quality of universities and research laboratories and the level and nature of support for research can be an important determinant in the migration decision and destination. For entrepreneurs, the climate for innovation and for business start-ups and self-employment is likely to be important, including issues of capital availability and market access. For overseas students, the quality of the higher education system is critical. To increase inflows of talented labour, policy should therefore act to reinforce these factors of attraction for specific target groups as well as to act on quality of life and "people climate" in more general terms.

Exploiting universities and research laboratories

Domestic universities and research laboratories can be used as a lever to attract post-graduate students and researchers, but exploiting this potential requires the right incentives and regulatory conditions in the public sector. Efforts are needed to improve the overall attractiveness of universities and public research laboratories to research staff. Of particular importance are salary levels and the quality of the research infrastructure. Increased flexibility in salaries systems is likely to help countries and regions to compete for top researchers. Awards, research grants and top-up salaries can also be important in attracting overseas researchers. This suggests a policy focus on building or maintaining research "centres of excellence". A number of countries have also introduced "repatriation" schemes for post-doctoral and older researchers. These work by offering fellowships, scholarships, supplementary funding or creation of posts to help returnees to re-integrate into scientific institutions in their origin countries. Competition for excellent researchers is often national and sometimes international, such that the top universities in any country or region need to set their benchmarks against world-class universities, including top United States universities, when seeking to retain world-class status in given fields.

To attract younger researchers, supply-side work on the quality of graduate education is critical. Indeed, the quality of graduate education is seen as one of the key reasons for the attractiveness of the United States system. The introduction of Integrative Graduate Education and Research Traineeship (IGERT) programmes offering stipend support to graduate students to engage in research in emerging multidisciplinary areas in science and engineering has contributed to this development. The number of PhD places available and access to funding for PhDs are also important influences on the decisions of young researchers to enter the university system. However, there is also room for work on the demand side, *i.e.* through improved marketing of opportunities to overseas students, especially post-graduates. A stronger educational marketing effort in the growing Asian market would be appropriate for many countries and regions, given the large and growing demand from Asia. In addition to this, some countries are having success in providing direct grants and tax incentives for emigrant students and researchers to return home to work in domestic universities. This is also a way of tapping into overseas knowledge through attracting talented people.

As well as attracting both young and established researchers into universities and public research laboratories, it is also important to foster researcher mobility between public institutions and industry in order to help knowledge to flow out of the universities and into firms ready to commercialise innovation. One method of achieving this is to provide

 GLOBAL KNOWLEDGE FLOWS AND ECONOMIC DEVELOPMENT – ISBN 92-64-10766-5 – © OECD 2004

academics with flexible employment arrangements that allow them to work part-time or temporarily in industry. Another way of increasing researcher mobility is to allow researchers working in industry to take fixed-term employment in universities, before in most cases returning to their original job. Public programmes can be developed to encourage such movement by providing a trusted framework for moves. To date, mobility between universities and research laboratories has tended to be weighted towards younger researchers. Policy might extend its impact by enlarging marketing and delivery approaches to include older and more experienced researchers.

Initiatives for renewal of the public research system have been undertaken in many OECD countries: graduate education has been reformed, funding for PhD training has been increased and reforms in the overall structure of the public sector, such as changes in tenure structure, have been undertaken. Similar efforts are recommended in other countries.

Adjustments to immigration rules and taxes

A number of countries eased their immigration rules in the 1990s to encourage the highly educated to work in industry. In particular, in the United States, the science and engineering workforce includes substantial foreign-born (often US-educated) people, and in some engineering and computer science fields foreign-born people exceed one-third of the total workforce. A number of visa classes are used to facilitate the temporary move to the USA of highly educated personnel and there is discussion at the US Congress about liberalising spouse work permit rules.

Reducing the tax burden on foreign experts and highly skilled workers can also be used to attract talented labour to a country. Denmark, the Netherlands and Belgium have each passed laws to alleviate the tax burden on foreign experts and highly skilled workers. In Quebec, the government is offering five-year income tax holidays (credits) to attract foreign academics in IT, engineering, health science and finance to take employment in the province's universities. In 2001, Sweden adopted similar policies for highly skilled workers who live in Sweden for less than five years. Nonetheless, the costs and benefits of such tax-based approaches should be examined closely.

Learning models from Scotland

Talent Scotland

The Talent Scotland programme has been in operation since 2001. It aims to raise the international profile of Scotland and its electronics companies in order to attract the world's finest engineers. The principal problem addressed is that most electronics companies in Scotland cannot recruit sufficient skilled engineers at the appropriate pace, largely as a result of information deficiency, characterised by low awareness of the strength of the company base, technologies, opportunities and quality of life in Scotland. The risk averse nature of individuals and organisations also contributes to recruitment problems in the electronics sector.

The aims of Talent Scotland are therefore to:

- Profile Scotland and its electronic technology companies.
- Target the world's finest experienced and recently graduated engineers and technologists to create a pool of talent that Scottish firms can draw on.
- Encourage the target audience to live and work in Scotland.
- Ensure engineers are kept fully up-to-date with opportunities in Scotland.
- Complement recruitment activities undertaken by companies, intermediaries and industry organisations.

The website *www.talentscotland.com* is the main communication route for the programme. The site includes general information on opportunities in Scotland to arouse the attention, interest and desire of engineers to look further. The web also allows engineers to register in three categories: *i)* general interest, *ii)* a job seeker posting their CV on the site, *iii)* a direct applicant to companies listed on the site. Around 7000 people had registered on the site in the fourth quarter of 2000. Approximately 1400 CVs are posted and approximately 95 companies are profiled. Some 150 job vacancies have also been posted, although the level has not been as high as originally anticipated because of the global downturn in the electronics sector, and more than 50% filled. Approximately 70% of the people registering on the site are not from Scotland. Major source countries include France, Germany, India, Pakistan and the United States.

The Talent Scotland initiative helps Scottish firms to hire senior engineers quickly from overseas where there are skill gaps that cannot be met by the existing Scottish workforce. The model is currently being tested in the electronics sector, but could be expanded to other key sectors in the Scottish economy.

Learning models from other OECD countries

Immigrant entrepreneurship in Silicon Valley

As one of the most prosperous regions in the United States, the San Francisco Bay area has become a focus for migration both from within the USA and from overseas, especially from Asia. It has one of the highest concentrations of immigrants of any US region. Today, immigrant workers, mainly Asian, constitute one-third of Silicon Valley's workforce. Many of the early Asian migrants in the late 1950s and 1960s started their own businesses because of their professional isolation. These Asian immigrants also established social networks, initially to celebrate holidays and family events with others who share language, culture and background. In time, professional and business associations evolved from these social networks. Today these organisations are among the most vibrant and professionally active in the Valley – facilitating professional networking, information exchange and mentoring. They have also helped build a two-way bridge between Silicon Valley and the immigrants' home countries. This helps both sides to locate foreign partners quickly and to manage complex business relationships across cultural and language barriers. The importance of both immigration and immigrant based networks to entrepreneurship in this dynamic region should be noted by policy makers and encouraged in other contexts.

The New Zealand research and education sector

The New Zealand education sector has made significant internationalisation and innovation advances over recent years, including:

- The creation of research consortia between universities, Crown Research Institutes and the private sector.
- The introduction of explicit innovation and commercialisation processes, especially:
  - ❖ Uni Services at the University of Auckland involving a proactive innovation search activity across university faculties and institutes and a commercialisation manager.
  - ❖ The Centre for Innovation at the University of Otago, which is a bespoke wet lab facility for the biotechnology sector, incorporating a new innovation search and management activity, a business incubator and corporate laboratory research suites where major companies can locate research staff to work on joint projects with university staff.
- The emergence of international student education as a major economic market for New Zealand.
- The introduction of performance-based research funding for tertiary institutions. Some 25% of the membership of the 11 funding panels will be from overseas. This is part of a reform of the tertiary sector to build regional and national centres of specialisation related to local industries, and eliminate excessive duplication which emerged under the previous enrolment-driven system.

This model provides an example of a comprehensive approach to improving the quality of the domestic higher education sector, both to support innovation and to attract overseas researchers and students.

Fast track immigration processes in New Zealand

Inward migration to New Zealand currently stands at record levels of around 50 000 per annum. To assist, companies of standing in key industries may accredit themselves to the New Zealand Immigration Service, enabling them to offer positions in the confidence that residency will be granted if agreed conditions are met. This scheme is working well for companies accredited to it.

The Korean electronics industry

In Korea, Samsung became the leading global chip producer in 1993, only a decade after launching its first memory chip product. Samsung and the other Korean firms entered the semiconductor industry through their own internal efforts and by leveraging the most advanced product and process technologies available in Japan, Europe and the US. Korea matched this investment with a large investment in technical training and by setting up technical and engineering institutes, which built Korea's absorptive capacity for advanced technologies. Global industry connectedness was achieved through diaspora and venture financing strategies. Many Silicon Valley semiconductor firms employed US-trained Korean-American engineers. These people were targeted to provide the scientific and engineering leaders that Korea needed to build its own semiconductor businesses. This example demonstrates the important role that attraction of returning migrants can play in domestic economic development.

Research centre funding in Ireland

In Ireland two government-administered schemes will distribute about 1.2 billion euros to advanced research centres by 2007. The first, the Programme for Research in Third Level Institutions (PRTLI), administered by the Higher Education Authority, was started in 1999 and has so far awarded over 600 million euros of research funding. A second scheme, Science Foundation Ireland (SFI), was launched in 2000 by the Department of Enterprise, Trade and Employment, and will give out 635 million euros in grants by 2007. The SFI grants will be made on merit review by top scientists and will target two fields: biotechnology and information and communications technologies. The schemes aim to attract researchers of any nationality, including Irish scientists currently working abroad. As an example, the newly established Programme for Human Genomics, funded with 45 million euros of PRTLI grant funding will require 40 graduate students, 40 post doctoral researchers and 20 academic staff. It is expected that PRTLI-funded centres will become self-sufficient, relying on external grants. Attracting grants will however, require Irish universities to reach critical size as research institutions.

The Singapore National Science and Technology Board

To attract foreign R&D labour, the Singapore government, through the National Science and Technology Board, has focused its efforts on improving and expanding investment in higher education and research facilities as well as creating centres of excellence. An example is the Institute of Molecular and Cell Biology (IMCB). Singapore is trying to develop a critical mass of skills in targeted areas. New centres include the Institute of Molecular Agrobiology (1995), the Bioinformatics Center (1996), the Institute of Materials Research and Engineering (1997), the Cancer Therapeutics Research Group (2000) and the Singapore Synchrotron Light Source (2001). To staff these institutes, the government is recruiting top talent from abroad. It is therefore putting more money into fellowships for new post-doctorates and post-masters. The country is targeting young graduates in mainland China and hopes that once they have completed their two-year fellowships that they will move onto industry. Like Scotland's Talent Scotland programme, Singapore has established an internet based job matching service that links up foreign faculty with firms and institutes in Singapore. Another programme assists companies in the recruitment of experienced foreign researchers, providing 50% funding for recruitment by helping with relocation costs, salaries and housing allowances for up to two years. In addition, the Temasek Professorships are used to target individuals in areas where Singapore wants to develop and to hire them to head or start laboratories. These areas include biological sciences, mathematics, semiconductors and data storage. Singapore plans to recruit between 20 and 30 Temasek professors for between 3 and 5 years and has allocated a budget of US$89 million. Recipients of the professorships, selected by a steering committee, will be required to spend only 50% of their time in Singapore.

The establishment of centres of excellence in research, financing of research posts, international Internet-based job matching, assistance to companies in recruiting overseas staff and creation of new professorships have all been used in seeking to attract talented labour and boost the science and technology base. Similar initiatives could be promoted in other countries and regions.

Recruitment of overseas students in Canada

Canada has introduced a range of programmes to attract undergraduate and postgraduate students from overseas. Financial incentives are being provided to students registering on graduate studies programmes. The number of Masters and Doctoral fellowships and scholarships awarded by the federal granting councils is being doubled. A world class scholarship programme is being created of the same prestige and scope as the Rhodes

Scholarships. A co-ordinated international student recruitment strategy is being supported by government and led by Canadian universities. Changes are being implemented to immigration policies and procedures to facilitate the retention of international students. A co-operative research programme is being established to support post-graduate students and, in special circumstances, undergraduates, wishing to combine formal academic training with extensive applied research experience in a work setting. In support of the repatriation of Canadian postdoctoral researchers, the Canadian Institute for Health Research offers a supplementary year of funding to Canadians and permanent residents who are recipients of either the Japan Society for the Promotion of Science (JSPS) Postdoctoral Fellowships for Foreign Researchers or Wellcome Trust/CIHR Postdoctoral Fellowships. Taken together, these initiatives could have an important impact on the number of students attracted from overseas.

Germany and Italy: Research awards

In Germany, the Humboldt Foundation and the Federal Ministry of Education sponsor a 22 million euro research award called the "Sofja Kovalevskaja-Preis" to help young scientists from overseas and expatriate German scientists to carry out research in Germany for a period of 3 years. The ministry for Research and Education (BMBF) has also launched a new programme in 2001 to attract the return migration of German researchers overseas. Similarly, Italy has recently introduced the "Reverse Brain Drain Project", aimed both at attracting foreign professors and scientists and at facilitating the return of Italian scholars abroad. In 2002, the Italian government funded additional positions worth 20 million euros. The use of research awards, prizes and funding of new positions are key tools that can be used for the attraction of talented researchers from abroad.

Building innovation systems

The final section of this synthesis aims to underline the point made a number of times in the book that it is important to build up national and regional innovation systems in order to create the pre-conditions for successful participation in global knowledge flows. It has been argued that a country or region's success in accessing global knowledge flows, and in turn increasing innovation and economic growth, will largely be a function of the strength of its innovation system. When policy initiatives are put into place, they should therefore set out to plug gaps in or enhance the infrastructures, institutions and networks in the innovation system, recognising the interactions between initiatives and their contribution to strengthening the overall innovation system.

 GLOBAL KNOWLEDGE FLOWS AND ECONOMIC DEVELOPMENT – ISBN 92-64-10766-5 – © OECD 2004

Adopting the innovation system perspective means recognising that firms, particularly SMEs, do not innovate in isolation but that innovation results from interactions with other firms (particularly customers, suppliers and collaborators), the knowledge infrastructure (particularly public research institutes and universities), and appropriate framework conditions (including regulations on intellectual property, the social and cultural context and the institutional framework). The innovation system concept is based on the idea that the overall innovation performance of an economy depends to a large extent on how far individual firms manage to utilise experience and knowledge in other firms and research organisations in their own innovation processes.

Different innovation system models have been discussed, including the Institutional Regional Innovation System (IRIS) and the Entrepreneurial Regional Innovation System (ERIS). In the former, the public sector plays an important role in developing institutions for knowledge generation, skills development and financing flows of knowledge from exploration to exploitation. In the best performing regions in the United States the market system appears to do this job well, hence the ERIS model is appropriate. But many regions in OECD countries have a weaker venture capital sector and fewer knowledge networks and the market is therefore less able to link together knowledge generators, human capital and finance. Public support is therefore often needed to build strong domestic innovation systems, resembling more closely the IRIS model.

Whichever model is pursued, a key issue is to avoid "lock-in" problems, which can occur if a system becomes inward looking and caught in a particular innovation segment. Part of the answer to avoiding lock-ins lies in accessing external knowledge through innovative co-operation and interaction between domestic firms and world-class national and international competence centres and innovation systems. This suggests taking a multi-level approach to innovation systems, recognising that different forms of knowledge must be accessed from different parts of the knowledge infrastructure (*i.e.* different organisation types) and at different spatial scales (*i.e.* local, regional, national, international). In addition, intermediary technology transfer agencies can often act as conduits to bring appropriate knowledge into the SME sector from elsewhere in the innovation system and abroad and thus counteract the potential lock-in problems.

Such intermediary technology transfer agencies often act as "broker" institutes between universities and research organisations and industry. They should attempt to fulfil the following functions:

- Strengthen the working of innovation systems by stimulating contact between firms and knowledge sources.

- Contribute to changing the innovative behaviour of firms in such a way that firms increase their external relations.
- Play an active role in identifying innovation barriers in firms and stimulating less innovative firms to start a learning process to become more innovative.
- Promote the formation of networks and interactive learning among firms and knowledge organisations.

The cases of Baden Württemberg in Germany and Finland suggest some learning points for countries and regions seeking to strengthen their domestic innovation systems (Boxes 7.1 and 7.2).

Box 7.1. **Baden Württemberg's Institutional Regional Innovation System**

The region of Baden Württemberg in Germany provides a good model of an Institutional Regional Innovation System (IRIS), although one that is probably better suited for incremental rather than disruptive technological change. Baden Württemberg has strong publicly-supported innovation centres that both generate knowledge and encourage its exploitation, performing a "boundary-crossing" role helping to forge the regional innovation system. These include the 14 Fraunhofer Institutes, which conduct applied research for larger firms, and the 300 Steinbeis Foundation transfer centres based in HEIs and Innovation Centres, which work largely with SMEs. The Fraunhofer Institutes conduct publicly-subsidised, industry-funded research to solve technological or managerial problems, assess technologies and conduct foresight activities. They make the bridge between the basic research carried out by the 14 Max Planck Institutes and the 10 universities in the region to the commercial application requirements of firms. The Steinbeis Foundation institutes provide a similar subsidised consultancy function for SMEs. However, neither the Fraunhofer nor the Steinbeis Foundation institutes engage directly in spinout activities, which tend to be promoted by innovation centres or incubators in science parks located close to universities.

The key to the success of an IRIS or public-led approach to innovation system development is to build linkages between the research results generated in universities and public research laboratories and their commercialisation in domestic companies. It is also important to integrate the domestic innovation system with overseas suppliers, customers and strategic partners in order to access global knowledge flows. Creation of intermediary

Box 7.2. **Finland: An open innovation system**

Finland's approach to strengthening its national and regional innovation systems is another good model of an IRIS, but the main reason for highlighting it here is because of its strong emphasis on external as well as internal linkages and its openness to global knowledge flows. Innovation policy in Finland is strongly embedded at the top governmental level through the Science and Technology Policy Council, which ensures that national and regional initiatives are well co-ordinated between the various government ministries. The innovation systems are built on key endogenous strengths in terms of good quality and extensive research and education, a competent workforce and good infrastructures. These endogenous strengths enable Finland to attract and retain knowledge-intensive firms, research projects and researchers and participate in global knowledge "pipelines" and have driven Finland's success in the development and use of high technology and its resulting increases in exports.

Two further factors are critical to Finland's approach. Firstly, regional Centres of Expertise support regional specialisation and inter-regional co-operation by identifying regional strengths and supporting the emergence of knowledge-intensive products, services, enterprises and jobs in each region. Their key role is to promote networking between industry, local government, technology centres, universities, polytechnics, research institutes and other branches of public administration. However, Finnish regional innovation systems are not closed to national and international links. Instead firms, universities and other players are encouraged to engage in global innovation networks and connect with the best centres of knowledge and research in the world. Secondly, universities are used to stimulate the inward transfer of knowledge from overseas and its dissemination to domestic firms, both by co-operating with firms that are research active and by extending tailored support services to disseminate knowledge to SMEs that are not usually involved in R&D activities themselves.

technology transfer agencies sitting between the universities and research laboratories undertaking basic research and the firms commercialising innovations is likely to be critical to the success of such strategies. Scottish Enterprise has recently undertaken an important new initiative to create three Intermediate Technology Institutes (ITIs), described in Box 7.3 below, whilst the Proof of Concept Fund and the Alba Centre, cited above, are also important mechanisms to bridge the gap between knowledge generation and knowledge exploitation.

Box 7.3. Intermediate Technology Institutes: Lessons from Scotland for building innovation systems

Launched in 2003, the Intermediate Technology Institutes (ITIs), see also Appendix 1, aim to accelerate the process of commercialising R&D in Scotland. Three Institutes are being established initially, with funding of some £450 million over 10 years, in the fields of energy research, life sciences and communications technology. The ITIs are managed by an executive team recruited from industry with the assistance of a scientific advisory board (including leading academics) and a market advisory board (including representatives of major Scottish and global companies and SMEs).

The rationale for setting up the ITIs is that Scotland has a dearth of high growth, technology based companies and is not doing as well as it would like in commercialising the technological research that it undertakes. Thus, the numbers of patents and scientific papers produced in Scotland is high, but corporate research expenditure is below the UK average and there are relatively few births of new technology based firms. This partly reflects a lack of existing technology-intensive firms to take on research created by Scottish universities and research laboratories. It also partly reflects problems in Scottish universities and research institutions in packaging innovations into products that can be exploited by companies.

The ITIs will therefore channel money into pre-competitive research. The process will start with identification of areas where technology needs developing in the given priority sector. Market opportunities will be identified by members of the Institute and the advisory boards, which will include potential users. Then, in a similar way to a corporate R&D laboratory, the Institutes will commission groups of researchers to produce the required technology in a series of research packages leading to exploitable products. The Institutes will be free to source the research either in Scotland or overseas using a system of expressions of interest, although Scottish universities are likely to be the major source of research. The aim is for this research then to be exploited in new or existing companies largely in Scotland. Since the Intellectual Property Rights and licensing fees will normally remain with the ITI, they are expected to generate own-revenues towards the end of the initial funding period agreed.

One of the strengths of the ITIs is that they will not be part of the university system itself, but instead will play an intermediary, boundary-crossing function between university research and commercialisation. Thus, the ITIs will locate near to universities but will be not sited on any individual university campus. Partnerships with universities will nonetheless be encouraged because there is a clear synergy between the pre-competitive research supported by the ITIs and the basic research undertaken by universities. Another strength of this model is that the Institutes should be able to achieve their goals at lower cost than by setting up major research laboratories, such as the Frauenhofer institutes in Germany, because they will make use of existing research infrastructure.

Experience from other intermediary technology transfer agencies (such as those described in Korea, Singapore, Taiwan and Baden Württemberg) suggests that for such intermediary agencies to be successful they need to:

- Develop partnerships with foreign affiliates to draw-in knowledge from an important group of players, in addition to that available in domestic universities and research laboratories.
- Focus on collaborating with firms and knowledge sources that can take near-market technologies to market. This is likely to have a greater impact on domestic innovation than simply translating basic research into technologies for licensing, whilst direct commercialisation could also build the number of new high technology start-ups.
- Build a critical mass of receptive, technology-advanced SMEs, able to commercialise technologies developed through intermediary agency contracts.
- Avoid the potential "lock-in" problems that can arise when innovation systems become focused on a particular technological trajectory and limited set of users. Methods of guarding against lock-in include competitive tendering for contracts and licensing, spinout and sale of core IPR arising from contracts and intermediary agency staff with the "absorptive capacity" to identify external innovations that could be exploited domestically.
- Seek support from a large proportion of the relevant domestic institutions and firms and build connectedness with other global centres of knowledge.

ISBN 92-64-10766-5
Global Knowledge Flows and Economic Development

APPENDIX 1

Global knowledge flows The issues for Scotland: a Scottish enterprise perspective

by

Martin Wight, Ross Brown and Michael Cannon
(Scottish Enterprise, United Kingdom)

Strategic context

"Scotland: Global Connections Strategy" sets out Scotland's strategic direction for taking advantage of the opportunities in the knowledge economy and ensuring that Scotland is a globally integrated economy. The two overarching objectives of the strategy are:

- "helping Scotland realise value by attracting knowledge from overseas";
- "helping Scottish knowledge generate value abroad for Scotland";

leading to the shorthand description of the strategy as focussing on Knowledge-in and Knowledge-out.

This paper sets out, as we embark on this partnership study with the OECD, the key issues identified, from a Scottish Enterprise perspective, in implementing the strategy through three important channels for achieving these objectives:

- Promoting cross-border alliances and activity by Scottish SMEs.
- Stimulating knowledge transfers from inward investment in Scotland.
- Attracting talented labour from other countries.

Before focussing on the issues under each of the three channels, it is worth setting the Global Connections Strategy in its wider Scottish context. It provides one strand of an overall approach set out in Scotland's economic development vision, "A Smart Successful Scotland", alongside complementary strands on Growing Business and Learning and Skills.

While the central thrust of the Global Connections Strategy is to enhance the Scottish economy's level of international activity in the context of increasing globalisation, its objectives, and the three streams of interest in this project, will be important means of meeting Scotland's broader economic challenges.

The overall challenge set out in "A Smart Successful Scotland" is to close the productivity gap between Scotland and the leading OECD economies. An analysis of GDP per head over the period 1995 to 1999 estimated that Scotland fell into the third quartile of OECD economies. The issues being considered in this project offer considerable potential in addressing some of the key components that can contribute to closing Scotland's productivity gap and boosting our economic growth rate.

- Increasing skill levels through attracting talent and the benefits for indigenous talent through stronger cross-border alliances and knowledge transfers from FDI.
- Increasing entrepreneurship through the same mechanisms. The Global Entrepreneurship Monitor ranked Scotland in the lowest of the three bands of entrepreneurship activity of 31 nations reviewed, alongside most other European countries, but below comparable small European economies such as Finland and Ireland.
- Increasing levels of innovation and R&D, particularly among indigenous business. This is among the highest priorities in transforming the Scottish economy and an area where the three channels for increasing international knowledge flows will play a crucial part. Scotland's total R&D expenditure as a proportion of GDP places it in the third quartile of OECD economies (compared to the UK's position in the 2nd quartile). Measures of research output from Scotland's universities (*e.g.* scientific research papers in leading journals per capita) suggests a high level of academic research activity, but this is not reflected in levels of R&D in the corporate sector.

These challenges sit alongside that of increasing the international-orientation of Scotland's businesses as drivers for this project.

Promoting cross-border alliances

Building the capabilities and confidence of Scotland's SMEs to participate in overseas markets is a crucial challenge for the future of the Scottish economy. As suggested above, it is part of a greater challenge to build a more innovative, ambitious and productive indigenous business base.

Any assessment of Scotland's current performance in making global connections begins with the health warning around data deficiencies, especially around knowledge flows, alliances, networking and joint ventures, and trade in services. Addressing these deficiencies is in itself a challenge.

The headline figures on manufactured trade suggest Scotland remains a very open economy with 12.4% of UK manufactured exports in 1999. However, digging below the headlines suggests a number of key issues:

- Non-UK owned companies now produce over 70% of Scotland's manufactured exports.
- 10 major companies account for around half of Scotland's manufactured exports.
- The 1999 figures estimate Scotland's share of UK service exports at only 4.7%.

Data on other forms of internationalisation is limited, but a recent survey of Scottish exporting firms found that only 119 UK-owned firms reported some form of overseas presence – covering overseas representative offices, subsidiaries, manufacturing operations, joint ventures, licensing agreements and franchises.

The policy challenge is therefore around:

- Building a broader base of SMEs who are internationalising, particularly in priority sectors.
- Deepening the internationalisation of Scottish SMEs.

There is currently a range of fairly young policy initiatives to respond to these challenges:

Global Companies Development Programme

The "Global Companies Enquiry" established Scotland's weakness in growing global companies, relative to similarly sized economies. The key weaknesses included:

- Low level of external networks and alliances.
- Lack of outward looking vision.
- Limited pool of internationally experienced managers.

The Programme has worked intensively with a relatively small number of potentially global companies (the original strategic target was 16 additional global companies by 2005) addressing the barriers and building their capacity to operate globally. The emphasis is on:

- Strategic review.
- Action planning for deeper forms of internationalisation.
- Management development.

The programme is now part of the broader activities of Scottish Development International's (SDI) International Business Development team, whose focus in 2003 is on:

- A concerted marketing and selling effort to raise awareness of the benefits of business internationalisation in Corporate Scotland.
- Leading the Scottish Enterprise Network in the development of new initiatives beyond the strategic planning focus of the current Global Companies Development Programme (GCDP).
- Broadening and deepening its work with the academic sector through access to both the Global and pre-Global Companies Development Programme.
- Leading efforts to engage foreign economic development agencies to help Scottish companies internationalise.

Improved company "Overseas Partnering" process

Work is currently underway to build on the export based Market Access Programme to develop a tool for overseas partnering, particularly around knowledge and technology partnering. The design is based on the principle that activity will be primarily pro-active on behalf of specific Scottish organisations, rather than reactive to potential interests in partnering expressed by overseas entities. The former self-selects motivated Scottish companies who are demonstrating a willingness to devote their own resources to partnering. The latter offers a very slim chance of finding a partner in Scotland for whom the unsolicited opportunity represents a good strategic fit.

This approach raises a number of operational challenges including:

- How to build alliances that will provide access to knowledge about the partnering opportunities across the globe, from which Scottish clients can benefit.
- How to benchmark and identify targets in this relatively new activity.

STAR *Centre incubation services*

Four Scottish Technology and Research (STAR) Centres in the United States provide a range of service options for Scottish businesses wanting to develop links or a presence in the US:

- Incubator – short lease office space.
- Transit – "hot desk" office space.
- Virtual – a US address.
- Research – intelligence on the US marketplace.

The first centre was established in 1998 in San Jose, with centres in Virginia and Florida opened in 1999 and most recently in Houston in 2000. An evaluation of the Centres was completed in 2002.

The evaluation recorded use of the above services by 120 companies, with significant business impacts. Customers see the reduction of risk on entering a new market as the centres' key attribute. The evaluation suggests that the Centres should have "the opportunity to contribute to the knowledge transfer aims of Global Connections. This could be in the form of the search for contracts for Scottish researchers or for the commercial sharing of Intellectual Property developed in Scotland." It does, however, raise issues around:

- The need to build the capacity of the Centres for this expanded role.
- Awareness of the Centres in Scotland.
- Their alignment with sectoral priorities.

Innovation relay centre

Scottish Enterprise was involved in the development and evaluation of the European Technology Transfer Network, and the related Innovation Relay Centre network remains active in Scotland. Delivered through 4 geographically dispersed providers (including one in the Highlands and Islands Enterprise area), Innovation Relay Centre Scotland aims:

- To work with Scottish organisations to facilitate the transfer of technologies to and from the rest of Europe to meet their needs.
- To assist Scottish organisations to participate in EU Research and Technology Development (RTD) funding programmes within a network of 68 Europe-wide centres. The Centre, which is primarily targeting SMEs, assists companies to make alliances around:
  - Finding partners interested in purchasing or licensing technologies that a Scottish based organisation has developed.
  - Helping Scottish based organisations acquire new technologies that can benefit and enhance their activities.

Delivering all of these initiatives successfully will rest on a set of common issues:

- At home, what is the best model of client management? We are currently engaged in ensuring our internationalisation work is integrated into a broad, consistent approach to account management across all of our business support.
- How can we build capacity in the field to deal with these broader knowledge flows?
- How can we extract maximum benefit from those Scottish organisations that are internationalising effectively?
- How do we tackle cultural issues around risk aversion and an inward-looking mentality?
- How can we best move alliances with other economic development agencies from sharing best practice to actively exchanging partnering leads?

Knowledge transfer from inward investment

An overview of FDI and the Scottish economy

Over the last 50 years, Scotland has very successfully capitalised on investment growth and liberalisation trends in the world economy; and in particular managed to attract a significant share of greenfield electronics assembly and manufacturing investment coming to Europe. The main reasons behind this success are well documented, including access to the EU,

 GLOBAL KNOWLEDGE FLOWS AND ECONOMIC DEVELOPMENT – ISBN 92-64-10766-5 – © OECD 2004

relatively low wages, inexpensive property, public subsidies and a flexible, English speaking workforce. Importantly, Scotland was also the first part of the UK to attract inward investment in a systematic and targeted manner. Its attraction policy largely replicated that of Ireland, with which it competed most directly for many investment projects.

Typically, foreign-owned firms investing in Scotland, such as IBM, HP/Compaq and NCR, produced high volume, standardised electronics products destined for the wider European marketplace. Most foreign-owned subsidiaries have shown weak innovation tendencies, consequent on limited local R&D, and very restricted linkages (of all types) into the rest of the economy.

Past studies examining the embeddedness of Scottish FDI tended to focus purely on the quantitative nature of buyer-supplier relationships, often overlooking important technological inter-firm relationships and internationalisation effects which flow from FDI. Therefore, it is important to look beyond the sheer numbers of linkages and look at their intrinsic quality, especially as this relates to the long-term economic health of a region.

Indeed, some observers have noted that the contribution to regional economic development of different types of plants is changing and that some MNEs are becoming more embedded within local economies. For example, some recent inward investment into Scotland (*e.g.* Agilent, Cadence, IBM and NCR etc.) has been of much higher quality than the high volume manufacturing FDI that typified much of the post war period. These firms are qualitatively different from most foreign-owned electronics companies in Scotland. Given the importance of the sector, we shall now examine the nature of knowledge flows generated by inward investors using the electronics industry as a case study.

Knowledge flows in the Scottish electronics industry

Since the 1950s, electronics has been the central driver of FDI in Scotland. Well over half of all employment in electronics is accounted for by foreign-owned companies.

At present, there are several main weaknesses in Scotland's electronics industry limiting the cluster's overall cohesiveness and many of these weaknesses result from poor knowledge transfers. These are summarised in Table A.1.1 below.

A major survey found that material linkages by value, between Original Equipment Manufacturers (OEMs) and local suppliers, were as low as 12 percent. Linkages between MNEs and local suppliers tend to be limited to basic activities conducive to localised supply, such as basic parts manufacture (plastics, cables and metal work) and turnkey supply services. Plus, increasing

Table A.1.1. **The cluster's main characteristics**

Cluster characteristics	Electronics
Ownership structure	Mostly foreign-owned MNEs and first tier suppliers, few strong local firms
Local demand condition	Limited Scottish market, poor demand conditions
Nature of manufacturing	Mostly high volume and standardised
Type of supplier linkages	Mostly dependent, adversarial supplier relations, suppliers often highly dependent on one or two main customers
Level of supply base internationalisation	Very low levels of internationalisation
Technological capabilities	Some advances, R&D in a few plants, most Scottish suppliers have no meaningful technological capability
Impact on local economic development	Large amount of direct employment, high levels of exports, little indirect benefits such as supply chain development or technology transfer
Cluster type and dynamics	Potential cluster: weakly embedded, low levels of innovation, at risk from external corporate decisions

Source: Brown, 2002.

consolidation and external takeovers in the supply base also means that the most progressive suppliers are increasingly externally-owned.

Links between OEMs and local research organisations are also low, usually limited to the few affiliates with a genuine R&D capability. The lack of a strong supporting sector undermines opportunities for genuine close partnerships and reduces the scope for technology transfer. Undoubtedly there have been positive demonstration effects for local companies, but these are difficult to measure or assess. There have also been several instances of local suppliers using local supply relationships to piggyback into external markets with the same MNEs in overseas markets. However, the supply base still has low levels of overall internationalisation compared with other sectors such as oil and gas.

In sum, the electronics industry has a large direct impact on the Scottish economy in terms of employment, output and exports but little impact in terms of indirect effects, especially in relation to supplier linkages, technology transfer and supplier internationalisation. This, together with the volume manufacturing nature of much of the industry in Scotland, means that the overall impact of the sector is reduced and places the industry at threat from divestment and eventual plant closure.

In July 2001, a five-year action plan was been agreed for the electronics industry in Scotland. The action plan has been jointly drawn up by Electronics Scotland, the main industry body in Scotland, and Scottish Enterprise. The action plan has three main objectives: increasing R&D in the industry, improving the competitiveness of the supply chain and improving skill levels within the industry. Following consultation with industry, the action plan began to be implemented in September 2001.

Current knowledge transfer policies in Scotland

In the past, policy towards aiding knowledge transfer from FDI in Scotland has been incoherent, poorly resourced and *ad hoc*. In many ways this reflects the hands-off approach towards industrial policy in the UK as a whole, and the traditional approach towards FDI where attraction rather than management was the main objective. There have been very few studies examining the effectiveness of Scotland's supplier development activities and even less work examining how Scotland compares with other countries in this area. Published information on the scope of these activities is therefore limited and partial.

However, here we highlight some of Scotland's past and present systems for encouraging knowledge flows:

1. Information and Matchmaking

   The only systematic initiative to encourage knowledge flows in Scotland was the development by Scottish Enterprise of a Supplier Development Programme in 1987. This programme sought to provide support to Scottish suppliers in the electronics industry and focused on supporting plastic moulding and sheet metal companies. The programme financed supplier audits, performed by contracted consultants working to a set of standard performance criteria. Based on the consultant's reports a series of improvement measures were identified and monthly meetings were established to monitor progress. As a means of indirect evaluation, Scottish Enterprise also tracked the share of local purchasing in annual procurement surveys of the largest electronics OEMs. Following some apparent early success, this programme was gradually downgraded during the late 1990s and has now ceased. Information and matchmaking activities are now very *ad hoc* in Scotland and depend on the initiatives of various cluster teams in Scottish Enterprise.

2. Technology Upgrading

   No programme has explicitly been developed to improve the technological capabilities of local suppliers in Scotland or to develop links between inward investors and the Higher Education sector. Although very little effort has been made to enhance linkages between incoming firms and universities and local research bodies, links have however emerged. Where this has taken place it has occurred through random, bi-lateral linkages between universities and local inward investors. For example, NCR in Dundee has developed links with Dundee and Abertay Universities, in order to develop technologies for new automatic teller machines (ATMs).

3. Training

Scottish Enterprise and its predecessor, the Scottish Development Agency, have been quite active in the field of training activities with both inward investors and local suppliers. First, Scottish Enterprise was instrumental in developing supplier associations, such as the Supplier Base Forum, which undertook training sessions and networking events for Scotland's electronics sector. Second, inward investors receive assistance with training needs through a host of firm specific and general vocational training packages. Such bespoke training initiatives have been identified as one of the key strengths of Scotland's FDI aftercare programmes. Assistance with training costs given to newer inward investors is increasingly important, especially for service-sector firms who do not benefit substantially from aid schemes such as Regional Selective Assistance (RSA) which tends to favour projects involving substantial capital expenditure. Third, Scottish Enterprise has also been active in promoting local training centres at Further Education colleges, often in close collaboration with a number of major inward investors.

Future issues

- Building linkages between new investors and indigenous businesses and universities.
- Building more holistic approaches to the development of niche sectors.
- Creating a comprehensive supplier development programme.
- Creating links between FDI and the new Intermediate Technology Institutes, which are to be established in Scotland.
- Exploiting the potential for new company spin-out from staff and managers in inward investors.

Attracting talented labour from overseas

One of the most concerning indicators of Scotland's economic health is its static or gradually declining population. The early results from the 2001 Census of Population show a decline of 116 000 (or 2%) since 1981. Scotland has not benefited from the large-scale migration patterns that are transforming labour supply in most developed economies. Scotland's net migration has fluctuated between positive and negative over time, largely in line with Scotland's economic performance relative to the rest of the UK. However this fluctuation has been within a narrow band and for some time Scotland has demonstrated the lowest population churn through migration of any UK region.

These figures relate to a deeper economic problem, but also result in less flow of knowledge through labour movement than in most developed economies. Our focus here is on talented labour. While data deficiencies make analysis difficult, they do allow us to conclude that the proportions of highly skilled workers and graduates are higher in out-migration than in in-migration.

This is a particularly new area of Scottish Enterprise's activity and is currently focussed on specific clusters and occupations. The aspirations of Project Alba to establish Scotland as a centre for electronic design quickly identified the need to attract a critical mass of people with a high level of design skills to complement local talent, in particular graduate output.

The approach taken at Alba has focussed on selling the existence of a wide range of high level opportunities in microelectronics, optoelectronics and software in Scotland. It is now a partnership project with the industry, branded Talent Scotland, which helps to recruit for vacancies from a growing group of companies, located across Scotland.

In a smaller initiative, the Bio Dundee project has supported the local university's attraction of academic talent in the life sciences. Here the decisive factor has been the pull of existing world class research and researchers in the institution. Biotechnology is currently being considered as a target sector in broadening the focus of Talent Scotland.

Both projects also sell the lifestyle benefits of living and working in Scotland, particularly in relation to larger urban centres with their congestion costs. SE is currently exploring the non-labour market factors that help distinguish the location decision of mobile talent, and a significant factor would appear to be the actual and perceived diversity of a place's current population. Scotland's low level of churn would appear to set it at a disadvantage in this respect.

Boosting Scotland's diversity and overall attractiveness to talent raises a number of key issues:

- Is it possible to change the perception/image of Scotland in advance of actual change?
- Can we boost the significant number of overseas students studying in Scotland's universities and provide incentives for them to remain in Scotland after graduating?
- In which sectors and clusters do we have sufficiently "thick" labour markets to attract talent?
- Can we tap into Scotland's established migrant communities as a mechanism for attracting new talent?

- This area of activity is clearly influenced by the current regulatory framework. As in many developed economies, UK policy is developing rapidly and is increasingly designed with the attraction of talent as a focus.

Is Scotland making sufficient use of this more positive policy context? While migration policy is a matter reserved by the UK government, can Scottish policy makers offer a distinctive contribution:

- As a test-bed for new ideas, in for example entitlements of overseas students or attraction of entrepreneurs?
- By raising awareness among Scottish businesses of the new opportunities offered by relaxation of work permit regulation?
- By using improved intelligence on the Scottish labour market to encourage a regional approach to determining favoured occupations within the UK?

Aside from the regulatory framework, are there operational issues where Talent Scotland can learn from talent attraction activities in other countries?

- In targeting of both occupations and target markets (returner v newcomer)?
- In the communication medium or message?
- Can it benefit from deeper integration with other components of the developing "knowledge-in" agenda and Scotland's international networks *e.g.* SDI's Field offices, GlobalScot and British Council networks?

Endnote: Intermediate Technology Institutes

A final important part of the Scottish policy context is the development of Intermediate Technology Institutes. They have the potential to make a significant contribution across all three areas of the enquiry.

The Institutes are a mechanism for encouraging and supporting pre-competitive research in key market areas in which Scotland has strong economic and business development potential.

Each Institute will have a small staff of approximately 15 who will co-ordinate with the market and undertake programme management. It is intended that all research activity be commissioned through existing research groups in Scotland and where necessary elsewhere. The three institutes, which will focus on Life Sciences, Energy and Communications Technology &Digital Media, will operate along the following principles:

- Provide a strong market focused link between the academic and corporate bases in Scotland and across the globe.
- Attract leading global players and researchers to participate in the identification of future market opportunities in which Scotland can become world leading.

 GLOBAL KNOWLEDGE FLOWS AND ECONOMIC DEVELOPMENT – ISBN 92-64-10766-5 – © OECD 2004

- Generate market focused intellectual assets by commissioning research from leading researchers in existing research institutions in Scotland and world-wide.
- Create pre-competitive market focused technology platforms that will strengthen existing Scottish companies, and/or create new firms.
- Facilitate the movement of staff, skilled in both market understanding and technology, between academia and industry.
- Retain technology graduates for Scotland by providing opportunities for career development.
- Capitalise upon existing science and technology strengths to develop Scotland as a world-renowned centre of expertise in specific technology niches.
- Act as a proxy for Corporate R&D and significantly increase the levels of R&D in companies in Scotland.
- Stimulate high technology firms in Scotland to create their own demand for near-market development research.

The proposed operating model is premised on the basis that the Institutes' strategy and operation will be actively guided and supported by its members. The attraction of members with a broad global perspective on markets and new technology directions, as well as a local focus to ensure that propositions can therefore be effectively transferred into the Scottish economy, are an essential condition of the proposed construction of the Institutes.

General membership access will be set at a level to allow SMEs in Scotland to easily become involved, another membership category will involve a higher level of commitment which firms will be able to pay in kind or as a financial contribution.

The Institutes' contributions to Scotland's economy have been identified as the following:

- Substantially increase and sustain the birth rate of indigenous high value-add technology based companies leading to significant improvements in productivity gains.
- Help with the move towards recruiting higher value-added FDI projects to Scotland.
- Substantially increase the level of exchange between academia and the corporate sector in Scotland, helping develop not just a transfer of skills and increased corporate R&D but also a more commercial and entrepreneurial culture and ethos in the Scottish academic science base.

- Strengthen and support the realisation of the commercial potential of the Scottish science base, leading to direct and significant productivity and competitiveness gains for existing companies through "hard" technology transfer, the transfer of skills, and the significant growth in the number of commercially focused researchers.
- Further establish and connect Scotland into key overseas markets and highlight Scotland as an important centre for specific technologies and as a location for foreign direct investment, portfolio direct investment and mobile skilled labour.
- Create a sustainable flow of market relevant technology companies that will attract more local and international venture capital leading to additional economic impacts.
- Reinforce and develop Scotland's key industrial clusters.
- Significantly increase the retention of graduate and professional skills in Scotland.

References

Brown, R. (2002) The future of ICT industries in Scotland, in N. Hood, J. Peat, E. Peters and S. Young (eds) *Scotland in a Global Economy: The 2020 Vision*, Palgrave Macmillan Limited: Basingstoke.

ISBN 92-64-10766-5
Global Knowledge Flows and Economic Development

APPENDIX 2

Contact details

THE INTERNATIONAL EXPERT PANEL

Professor Bjørn T. Asheim

Department of Social and Economic Geography
University of Lund
Sölvegatan 13
SE-223 62 LUND
Sweden

Tel.: ++ 46 46 222 84 02
Fax: ++ 46 46 222 84 01
Email: Bjorn_T.Asheim@keg.lu.se

Mario Cervantes

Economist, Science and Technology Policy Division
Directorate for Science, Technology and Industry, OECD
2, rue André-Pascal
75775 Paris Cedex 16
France

Tel.: ++ 33 1 45 24 94 31
Fax: ++ 33 1 44 30 62 64
Email: mario.cervantes@oecd.org

Professor Philip Cooke

Director, Centre for Advanced Studies
Cardiff University
44-45 Park Place
Cardiff, CF10 3BB
United Kingdom

Tel.: ++ 44 2920 874 945
Fax: ++ 44 2920 874 994
Email: cookePN@cardiff.ac.uk

Paul Frater

Director, Strategic Direction, Industry New Zealand
22 The Terrace
PO Box 2878
Wellington
New Zealand

Tel.: ++ 64 449 66 298
Fax: ++ 64 449 55 081
Email: paul.frater@industrynz.govt.nz

Professor John de la Mothe

Canada Research Chair in Innovation Strategy
School of Management
University of Ottawa
Vanier Hall
275 Nicholas Street
Ottawa
Canada
KN1 6N5

Tel.: ++ 1 613 562 5800
Fax: ++ 1 613 562 5164
Email: delamothe@admin.uottawa.ca

MAIN OECD CONTACT

Dr Jonathan Potter

Senior Economist, LEED Programme
Directorate for Employment, Labour and Social Affairs, OECD
2, rue André-Pascal
75775 Paris Cedex 16
France

Tel.: ++ 33 1 45 24 89 77
Fax: ++ 33 1 45 24 16 68
Email: jonathan.potter@oecd.org

MAIN SCOTTISH ENTERPRISE CONTACT

Martin Wight

Scottish Enterprise, 150 Broomielaw,
Atlantic Quay
Glasgow
G2 8LU
United Kingdom
Tel.: ++ 44 141 248 2700
Fax: ++ 44 141 221 3217
Email: Martin.Wight@scotent.co.uk

Questionnaire on the quality of OECD publications

We would like to ensure that our publications meet your requirements in terms of presentation and editorial content. We would welcome your feedback and any comments you may have for improvement. Please take a few minutes to complete the following questionnaire. Answers should be given on a scale of 1 to 5 (1 = poor, 5 = excellent).

Fax or post your answer before 31 December 2004, and you will automatically be entered into the prize draw to **win a year's subscription to *OECD's Observer magazine*.***

A. Presentation and layout

1. What do you think about the presentation and layout in terms of the following:

	Poor	Adequate		Excellent	
Readability (font, typeface)	1	2	3	4	5
Organisation of the book	1	2	3	4	5
Statistical tables	1	2	3	4	5
Graphs	1	2	3	4	5

B. Printing and binding

2. What do you think about the quality of the printed edition in terms of the following:

Quality of the printing	1	2	3	4	5
Quality of the paper	1	2	3	4	5
Type of binding	1	2	3	4	5
Not relevant, I am using the e-book	❑				

3. Which delivery format do you prefer for publications in general?

Print ❑ CD ❑ E-book (PDF) via Internet ❑ Combination of formats ❑

C. Content

4. How accurate and up to date do you consider the content of this publication to be?

1 2 3 4 5

5. Are the chapter titles, headings and subheadings...

Clear	Yes ❑	No ❑
Meaningful	Yes ❑	No ❑

6. How do you rate the written style of the publication (*e.g.* language, syntax, grammar)?

1 2 3 4 5

D. General

7. Do you have any additional comments you would like to add about the publication?

....

....

....

Tell us who you are:

Name: .... **E-mail:** ....

Fax: ....

Which of the following describes you?

IGO ❑ NGO ❑ Self-employed ❑ Student ❑

Academic ❑ Government official ❑ Politician ❑ Private sector ❑

Thank you for completing the questionnaire. Please fax your answers to:
(33-1) 49 10 42 81 or mail it to the following address:
Questionnaire qualité PAC/PROD, Division des publications de l'OCDE
23, rue du Dôme – 92100 Boulogne Billancourt – France.

Title: Global Knowledge Flows and Economic Development

ISBN: 92-64-10766-5 **OECD Code (printed version):** 84 2004 01 1

* Please note: This offer is not open to OECD staff.

OECD PUBLICATIONS, 2, rue André-Pascal, 75775 PARIS CEDEX 16
PRINTED IN FRANCE
(84 2004 01 1 P 1) ISBN 92-64-10766-5 – No. 53455 2004